BIOTECHNOLOGY PROJECTS FOR YOUNG SCIENTISTS

Kenneth G. Rainis and
George Nassis

FRANKLIN WATTS
A Division of Grolier Publishing
New York ▪ London ▪ Hong Kong ▪ Sydney
Danbury, Connecticut

For my aunt Mary (Violet) Goff
a courageous and caring person

For my father Stergios Nassis
whose love for education has sent me down the path
of lifelong learning

NOTE TO READERS:
Throughout this book, some measurements are given only in metric or in English units. There are two reasons for this. In some cases, materials or equipment is sold according to either metric or English measurements. In other cases, it is important for you to know a measurement's value precisely. A degree of accuracy is lost when a converted measurement is rounded to the nearest significant figure.

Photographs ©: BioMedia Associates: 81 (Bruce Russell); Dr. Ronald Carter and Viola Freeman Art: 96, 100, 147, 148, 149, 150, 151, 152; Rainbow: 12 (Hank Morgan); Photo Researchers: 55 (E.R. Degginger), 94 (James King-Holmes/ICRF/SPL), 48 (Robert Noonan), 77 (Rosenfeld Images Ltd/SPL), 114 (Alvin Steffan); Tony Stone Images: 13 (David Joel); UPI/Corbis-Bettmann: 7; Ward's Natural Science Establishment: 15.
Illustrations created by Ken Rainis, Bob Italiano, George Stewart, and Claire Fontaine.

Visit Franklin Watts on the Internet at:
http://publishing.grolier.com

Library of Congress Cataloging-in-Publication Data

Rainis, Kenneth G.
    Biotechnology projects for young scientists /Kenneth G. Rainis and George Nassis.
        p.   cm. — (Projects for young scientists)
Includes bibliographical references and index.
Summary: Gives instructions for and explains the principles behind a variety of biotechnology experiments from the simple to the more difficult.
ISBN 0-531-11419-8 (lib. bdg.)   0-531-15913-2 (pbk.)
1. Biotechnology projects.   [1. Science projects.   2. Experiments.]
I. Nassis, George.   II. Title.   III. Series.
TP248.22R35   1998
660'.6.'078—dc21   97-20536
                                                                              CIP
                                                                              AC

C O N T E N T S

ACKNOWLEDGMENTS

This book could not have become a reality without the support and encouragement of our wives, Joan Rainis and Olga Nassis.

Our editor, Melissa Stewart, always supportive, was our champion and tether—making sure we did not rise above our intended audience.

We gratefully acknowledge the generosity, support, and encouragement of Kurt Gelke, president of WARD'S Natural Science Establishment. The critiques of individuals at WARD'S—Robert Iveson and Geoffrey Smith—helped make this a more accessible work for young scientists. Thanks to Gervase Pevarnik for his photographic expertise and Jeff Foti for botanical assistance. Stephanie Miller, Sean Hendrick, and Lou Lopez helped create the line art illustrations that appear in this book.

Special thanks goes to Dr. Ron Carter and Viola Freeman, ART of the Cytogenetics Laboratory, McMaster Hospitals, Hamilton, Ontario, Canada, for their assistance regarding human chromosome spread images for the karyotyping activities.

Michael Peres of Rochester Institute of Technology assisted in microtechnique photomicrography. Bruce J. Russell of BioMEDIA Associates provided special micro-life images.

WHAT
IS
BIOTECHNOLOGY?

Before you start doing biotechnology projects, it is important to know exactly what biotechnology is and why it is useful. The first part of the word—"bio"—means "living." The second part of the word—"technology"—means "the practical application of knowledge." Thus, biotechnology can be defined as the practical application of knowledge acquired from the study of living things.

Many of these applications are not new to us. People have been making food, medicines, and other products using the unique properties of living organisms for centuries. Ancient Egyptians knew how to make wine and beer by fermenting grapes and grains. For centuries, humans have taken advantage of the characteristics of living organisms to make cheese and bread. We have known that it is possible to create corn that grows larger ears and

Ancient Egyptians crushing grapes for wine

cows that produce more milk by selectively breeding species with particular combinations of traits.

What's new about biotechnology is our understanding. Ancient Egyptians and early farmers didn't really know why or how fermentation or selective breeding worked. They learned by trial and error. Making the best cheese, bread, or alcoholic beverages was considered more of an art than a science. Today, scientists understand what happens at each step of the complex biological and chemical processes.

These processes are carried out according to instructions provided by each cell's deoxyribonucleic acid (DNA). Once scientists understand the genetic code that regulates cell activity, they can look for ways to alter and manipulate cellular functions to manufacture specific products in a predictable and controllable way. This is called *bioengineering*. Bioengineers use the cells of microorganisms to break down toxic wastes and produce pharmaceutical products, such as insulin. They develop tests that can be used to diagnose genetic diseases or identify criminal suspects. Bioengineers can also alter plants to make them hardier and more productive.

THE PROJECTS IN
THIS BOOK

The projects in this book are designed to help you explore biotechnology. You will learn how cellular processes can be manipulated to achieve a desired effect. Some of the projects are easy, and some are difficult. Some are presented in step-by-step fashion, like a laboratory investigation, while others are simply ideas for you to develop on your own.

You can work through the projects systematically to learn more about the nature of genetic material and the role biotechnology plays in agriculture, medicine, industry, and the environment, or you can skip around and try only the projects that interest you most. Many of the experiments could be done to satisfy a class requirement or to enter in a science fair.

We have tried, wherever possible, to suggest materials that are readily available around the home or at the grocery store. The best source for science apparatus, microorganisms, and chemical reagents may be your sci-

SAFETY: THE MOST IMPORTANT PART OF ANY PROJECT

1. Be serious about science. A careless attitude can be dangerous to you and to others.

2. Read instructions carefully before starting any project outlined in this book. If you plan to expand on these activities, **discuss your experimental procedure with a science teacher or other knowledgeable adult.**

3. Keep your work area clean and organized. Never eat or drink while conducting experiments.

4. Respect **all** life forms. Never mishandle organisms or perform an experiment that will injure or harm them.

5. Wear protective personal equipment—safety goggles, protective gloves, and a smock or an apron—whenever you handle chemicals, heat substances, or perform experiments that could harm your skin or eyes.

6. Do not taste chemicals or chemical solutions. Avoid inhaling the vapors or fumes of any chemical or chemical solution.

7. Clean up chemical spills immediately. If you spill a chemical on your skin or clothing, rinse it off with plenty of water, and then tell a responsible adult.

8. Keep flammable and combustible materials away from heat sources and sparks.

9. **Never** operate a power tool without adult supervision.

10. **Always** wash your hands after conducting experiments. Dispose of contaminated waste properly.

ence teacher. You will find information about ordering equipment in the Materials Guide at the back of the book. The Technical Notes section provides in-depth information and guidance about growing microorganisms, preparing solutions, and performing other biotech techniques. The Resources section lists books and organizations that can provide background information and supply additional project ideas.

THE SCIENTIFIC METHOD

As you work, follow the same process that professional scientists use:

Observe → Guess → Test → Conclude.

Be sure to notice changes as your experiments proceed and record all data and observations in a lab notebook. Make a guess (hypothesis) about what might be causing the changes you observe. Test your hypothesis with an experiment and use your results to conclude whether your hypothesis is correct. If your data do not support your hypothesis, try another experiment. In some cases, you may want to use a *control* to make sure that outside factors are not influencing your experimental results.

Your powers of observation will provide the springboard to your investigations. At times, your investigations will be aided by serendipity—the art of finding something valuable when you are looking for something else. As the great scientist Louis Pasteur said, "Chance favors the prepared mind."

TECHNIQUES OF BIOTECHNOLOGY

This chapter provides instructions for building basic biotechnological equipment and describes a number of important techniques. Once you learn the fundamentals, you can move on to the projects presented in Chapters 2 through 7.

GEL ELECTROPHORESIS

Individual genes or DNA molecules are much too small to be seen directly, except with the most powerful microscopes. However, it is relatively easy to visualize proteins or short segments of DNA and RNA using a technique called *gel electrophoresis*. The word "electrophoresis" can be broken into two parts—"electro" is anything related to electrical energy, and "phoros" means "to carry across." The technique involves the separation and movement of

11

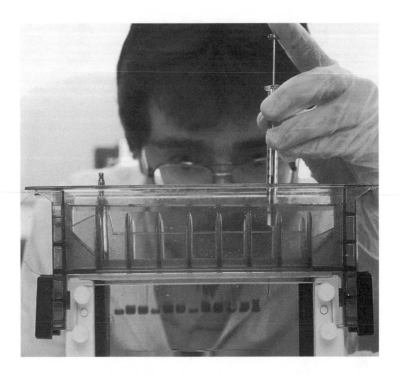

A scientist loading samples into a vertical
electrophoresis apparatus.

molecules through a gelatinlike material that has been
exposed to an electrical field. Visualizing DNA, RNA,
and proteins helps scientists understand how they function in our bodies.

An electrophoresis apparatus has five major components—the driving force (electrical current), the sample
(DNA, RNA, or protein), the support medium (agarose
gel), the buffer (liquid electrical conduction medium),
and the detecting system (stain).

Samples, such as DNA fragments, are typically placed
in the wells at one end of the gel slab. The gel is then
placed in the conductive buffer solution. Electric current

This scientist is examining a film made from an electrophoretic gel of a DNA strand. The components of the DNA have been separated, creating a banding pattern that resembles a bar code.

is applied at both ends of the electrophoresis chamber, creating an electric field within the gel. Because DNA is negatively charged, DNA fragments always migrate toward the positive pole of the apparatus—the end opposite the wells. Some samples move more quickly than others, depending on the size of the molecules.

After the gel is removed from the apparatus, it is stained with a dye so that the movement of the samples can be viewed. The separated molecules form a series of bands spread from one end of the gel to the other. This banding pattern looks something like a bar code.

Each band on a gel electrophoresis represents a specific identifiable fragment of DNA. Scientists and clinicians use electrophoresis to identify diseased tissue, characterize genetic dysfunctions, assess coronary risk, pinpoint cancer types, and read the *nucleotide base* sequence of a particular gene. You will learn more about DNA in Chapter 2.

You can purchase an electrophoresis apparatus from a science supply company, or you can make your own apparatus using the instructions outlined in the following procedure.

BUILDING AN ELECTROPHORESIS APPARATUS

What You Need	
Awl	2×3-inch hinged transparent plastic box
12-ounce rectangular Tupperware™ container with cover	3×5-inch clear plastic strip
Two $\frac{1}{8} \times 2$-inch stainless steel cotter pins	Side cutters
Permanent marker	Medium-grade sandpaper
Masking tape	2% agarose gel solution
Silicone sealer	Hot-water bath or microwave oven
Coping saw	

Caution: Do this project under adult supervision. Wear safety goggles and insulated gloves.

Using an awl, carefully puncture one hole about $\frac{1}{2}$ inch from each end of the rectangular Tupperware™ container.

An electrophoresis apparatus consists of a chamber
filled with buffer, a gel casting tray that holds
the agarose gel, and a power source. When an electric
current is passed through the chamber, samples
move through the agarose gel according to their
molecular weight.

Insert a cotter pin into each hole. About $\frac{1}{4}$ inch of each pin should remain outside of the container. See Figure 1. These pins will carry the driving force (electric current) needed to separate the molecules that make up a DNA fragment.

Designate the electrode on your right as negative, and the one on your left as positive. Label them (+) and (–) using a permanent marker and masking tape.

Seal the hole around each pin with silicone sealer. Allow the sealer to dry for about 24 hours, then fill the Tupperware container with water and make sure that it doesn't leak.

Before you can run a sample, you must make an agarose gel slab using a gel casting tray. To make the casting tray, remove the top of the 2 × 3-inch plastic box and carefully saw off both ends with a coping saw.

A plastic comb should be used to make wells in the slab. You can make a comb by cutting slots out of a piece of the clear plastic strip and then removing the pieces with side cutters. To do this, draw six to ten equal-sized "teeth" along one side of the plastic strip. Use a piece of medium-grade sandpaper to smooth out the edges of the comb.

To make the agarose gel slab, melt 2% agarose—purchased from a science supply company—in a hot-water bath or a microwave oven (1-minute cycles, low setting). (Instructions for preparing your own 2% agarose solution are provided in the Technical Notes at the back of the book.)

Tape both of the cut-off ends of the gel casting tray with masking tape. Be sure the tape forms a good seal, so that the tray can hold a liquid. Carefully pour the melted agarose into the casting tray to a depth of about $\frac{1}{4}$ inch. Place the well-forming comb approximately 1 cm from

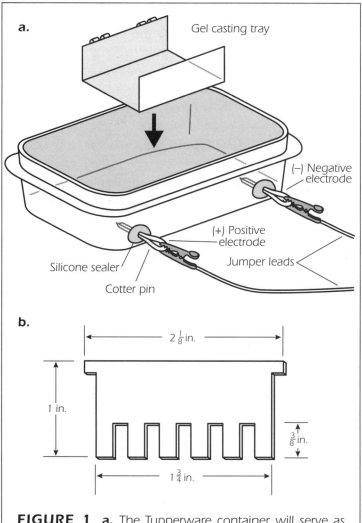

a. Gel casting tray

(−) Negative electrode

(+) Positive electrode

Jumper leads

Silicone sealer

Cotter pin

b.

$2\frac{1}{8}$ in.

1 in.

$\frac{3}{8}$ in.

$1\frac{3}{4}$ in.

FIGURE 1 **a.** The Tupperware container will serve as the chamber of the electrophoresis apparatus. Jumper leads with alligator clips are attached to the cotter pins. The gel casting tray is placed inside the chamber. **b.** After the agarose is prepared and poured into the gel casting tray, the comb is added to create wells for the samples.

the end of the tray.* Allow the gel to solidify for about 20 minutes. Be sure not to disturb the gel or comb during this period. When the slab begins to look opaque, carefully remove the comb. When the slab has solidified completely, remove the tape from each end of the casting tray, but do not remove the gel from the tray.

USING AN ELECTROPHORESIS APPARATUS

You can test your electrophoresis apparatus by running a gel containing several different food dyes. Like DNA, these dyes are negatively charged and will migrate toward the positive electrode (anode). Dye mixtures will separate into discrete bands based on their molecular size.

What You Need	
15 g of granulated sugar	Electrophoresis apparatus
30 mL of distilled water	
Small container	Diluted running buffer (TBE or sodium chloride solution)
4 vials with caps	
9 medicine droppers	Five 9-V batteries
Food dyes (red, green, yellow, blue)	Two 14-inch black and red jumper leads

*If you wish to use the apparatus to determine the molecular charge of unknown molecules, place the well-forming comb in the center of the tray. This will allow you to observe the samples' direction of migration.

Add 15 grams of granulated sugar to 30 mL of distilled water in a small container. Remove the caps from each of the four vials. Use a medicine dropper to add 5 drops of sugar solution to each vial. Using separate medicine droppers, add 6 to 8 drops of each dye to separate vials.

Using separate medicine droppers, dispense a minute drop of each prepared dye solution into separate wells in the gel. (The addition of the sugar to the dye increases the density of the solution, making it sink to the bottom of the wells in your gel.) Be careful not to puncture the sides or bottom of the wells while dispensing your samples.

Place the casting tray with the gel into the electrophoresis apparatus. Gently add diluted running buffer (1X concentration) to the apparatus until 1 to 2 mm of liquid cover the gel.* Instructions for preparing running buffer are provided in the Technical Notes at the back of this book.

Place the cover over the plastic Tupperware container. Arrange the five 9-volt batteries as shown in Figure 2 on the next page and connect this power source to the electrodes using the jumper leads. The red jumper lead should run between the positive electrode terminal on the chamber and the positive electrode on the battery pack. The black jumper lead should run between the negative electrodes on the chamber and battery pack. As soon as the battery pack is connected to the chamber, you should see bubbles forming along the electrodes in the chamber. These bubbles are the result of electrolytic decomposition of water.

Within several minutes, the samples will begin to

---

*Commercially available running buffer is generally sold as a 10X concentrated solution. Prior to use, it must be diluted with distilled water.

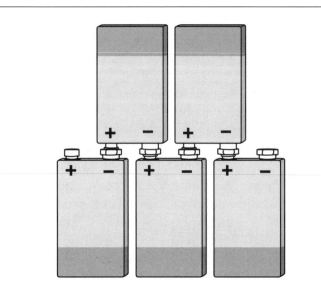

FIGURE 2 Assemble five 9-volt batteries as shown above and attach the alligator clip on one end of the red jumper to the positive electrode. The other end of the red jumper lead should be attached to the cotter pin designated as the positive electrode. The black jumper lead should connect the negative electrode of the chamber to that of the battery pack.

migrate toward the positive electrode. Disconnect the battery pack when the dye bands have fully separated and the smallest band has reached the end of the gel.* Make an accurate drawing of the gel in your notebook. Be sure to indicate the exact position of each band. Compare your drawing to Figure 3. Measure the distance each dye band migrated from the well or point of origin. Record this

*Although gels of DNA, RNA, and protein samples require staining, the sample run is this project does not. The food dyes readily bind to the gel and can be easily viewed.

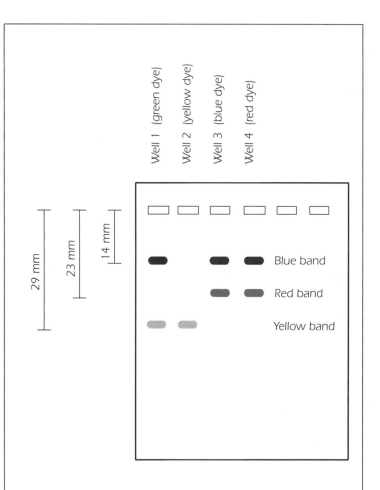

FIGURE 3 The bands that traveled 14 mm indicate the presence of blue. The bands that travelled 23 mm indicate the presence of red. The bands that travelled 29 mm indicate the presence of yellow. From this electrophoresis, it is clear that green dye has yellow and blue components; blue dye has some red components; and red dye has some blue components.

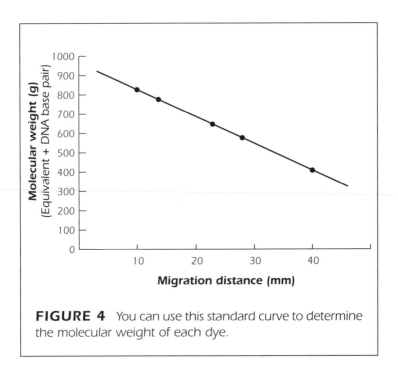

FIGURE 4 You can use this standard curve to determine the molecular weight of each dye.

information and the direction of migration in your lab notebook.

The molecular size of an unknown molecule can be determined by comparing its electrophoretic mobility on an agarose gel with that of a known sample. The standard curve shown in Figure 4 is a representation of migration distance versus molecular weight. Use the information in the graph to determine the molecular weights of the dyes you used.

Doing More

As an additional project, you may want to compare the direction of migration of negatively charged food dyes along with positively charged dyes such as 1% crystal violet or methyl green. Ask your science teacher to provide you with samples.

BUILDING A TRANSFER CHAMBER

A sterile environment is required when working with cultures of bacteria or plant tissues. A *culture* is a population of cells or organisms that have been deliberately grown on or within a solid or liquid medium. The sterile environment greatly reduces the chances that equipment, culture media, and living tissues will become contaminated by bacteria and other materials floating in the air. Transfer chambers are commercially available, but they cost thousands of dollars. You can construct your own transfer chamber using the instructions below as a guide.

What You Need	
Sturdy cardboard box approximately 3 × 2 × 3 foot	Heavy-duty aluminum foil
Scissors	Diluted household bleach (1 part bleach to 9 parts water) or rubbing alcohol
Heavy-duty clear plastic food wrap	
Duct tape	Fluorescent desk lamp
Paper towels	

Caution: Wear safety goggles, rubber gloves, and an apron or a smock. Bleach is corrosive. Do not use an incandescent light to illuminate the chamber. It could

23

melt or ignite the plastic wrap. Do not use the transfer chamber as a chemical fume hood.

Cut the box so that the front is 2 feet high, the back is 3 feet high, and the sides slope diagonally, as shown in Figure 5. Cut two 6-inch circular openings in the front of the box. Make sure that your hands fit through these openings. Line the entire inside surface of the box with heavy-duty aluminum foil. Cover the top of the box with heavy-duty plastic food wrap. Secure the plastic wrap with duct tape.

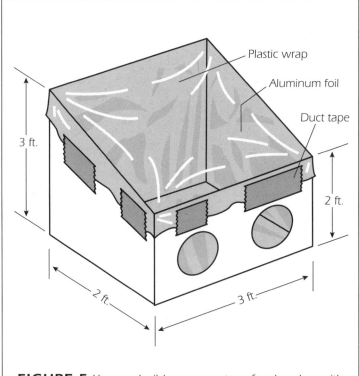

Plastic wrap

Aluminum foil

Duct tape

3 ft.

2 ft.

2 ft.

3 ft.

FIGURE 5 You can build your own transfer chamber with a carboard box, plastic wrap, duct tape, and aluminum foil.

To sterilize your transfer chamber, wipe down the inside surface of the box with a paper towel and diluted bleach. Be sure to resterilize the box before placing a new sample inside. Keep the transfer chamber in a well-lit area. If you need additional light, you may use a fluorescent desk lamp, but make sure the light is at least 6 inches away from the transfer chamber.

BIOREACTORS
(FOR ADVANCED YOUNG SCIENTISTS)

Fermentation is a chemical process that breaks down energy-rich organic materials. The process is carried out by microorganisms such as bacteria, molds, and yeasts. The best known fermentation process is the breakdown of plant carbohydrates into ethyl alcohol and carbon dioxide. Yeasts break down the carbohydrates in grains and grapes to make beer and wine. Fermentation is also essential in the production of bread, cheese, and yogurt. In some cases, fermentation makes foods inedible. For example, when milk ferments, it turns "sour."

Today, we don't have to rely on natural fermentation. We can use fermenters, or bioreactors, to manufacture alcoholic beverages, cheese, bread, solvents, and pharmaceutical products. In a *bioreactor*, living cells are mixed with nutrients and grown in a carefully controlled, sterile environment to make a specific product. The temperature, pressure, pH, and oxygen content must be maintained at just the right levels for the process to work correctly.

In the following projects, you will build a functional bioreactor and then use it to grow bacteria that will produce grease- and protein-cutting enzymes.

BUILDING A BIOREACTOR

What You Need	
Petroleum jelly	Three 3-inch lengths of vinyl air line tubing with $\frac{5}{16}$-inch outside diameter
3-hole rubber stopper (#8)	
Two 11-inch lengths of rigid plastic tubing with $\frac{3}{16}$-inch outside diameter	Screw compress clamp
	2 air filters
One 5-inch length of rigid plastic tubing with $\frac{3}{16}$-inch outside diameter	1-liter glass bottle or Erlenmeyer flask
	One 3-foot length of vinyl air line tubing with $\frac{5}{16}$-inch outside diameter
Aquarium air diffusion block	
10-cc syringe pipette without needle	Fish-tank air pump

Apply a small amount of petroleum jelly to each hole in the rubber stopper. Carefully insert the three pieces of rigid plastic tubing into the holes of the stopper. Attach the aquarium air diffusion block to the end of one 11-inch length of rigid plastic tubing. This tube will be referred to as the air inlet. The air diffusion block will provide the microorganisms with a constant supply of oxygen.

Attach one of the 3-inch pieces of vinyl air line tubing to the top of each piece of rigid tubing. See Figure 6. Attach a screw compress clamp to the other piece of 11-inch rigid plastic tubing. This tube will be referred to as the sampling port. The syringe pipette without needle

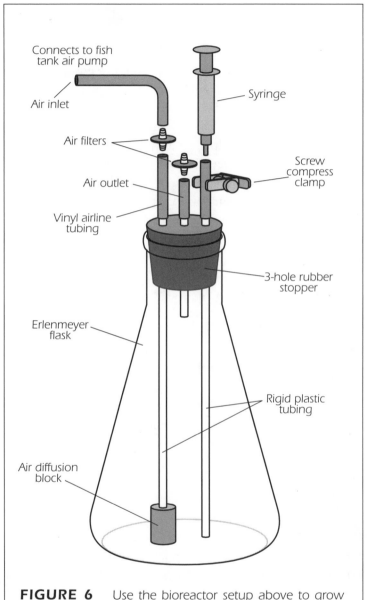

Connects to fish
tank air pump

Air inlet

Syringe

Air filters

Screw
compress
clamp

Air outlet

Vinyl airline
tubing

3-hole rubber
stopper

Erlenmeyer
flask

Rigid plastic
tubing

Air diffusion
block

FIGURE 6   Use the bioreactor setup above to grow samples of bacteria with specific characteristics.

will be placed on top. The 5-inch length of rigid tubing will serve as an air outlet.

Attach an air filter (available from science supply companies or at a store that sells wine-making equipment) to the air inlet and the air outlet tubes and place the entire apparatus into the 1-liter bottle or flask. The rubber stopper should seal the opening of the flask completely. The inlet filter will sterilize the air as it goes into the flask, and the outlet filter will prevent noxious aerosols from escaping. Run the 3-foot length of air line tubing from the air inlet to the fish-tank air pump. You are now ready to use your bioreactor.

MANUFACTURING MICROBES

Your bioreactor has most of the important features of modern commercial fermenters. You can use it to grow "drain bacteria." These microbes, which are available in freeze-dried form at most hardware or plumbing supply stores, produce enzymes that can break down the protein and fat deposits that can clog drains in sinks.

You will know that the reaction is going smoothly if the mixture becomes increasingly turbid (cloudy). This change occurs because the density of cells is continually increasing as cells grow and divide. If your solution does not become cloudy, your cells are dead, and you must begin again.

The growth of a bacteria population can be illustrated as a graph in which the increase in cell numbers is plotted against time. See Figure 7. There are four main phases in a growth curve: The lag phase, the growth phase, the stationary phase, and the death phase. The lag phase is a period of little or slow growth. During the growth phase, the bacterial cells grow rapidly. During the stationary phase, growth declines and eventually stops as nutrients

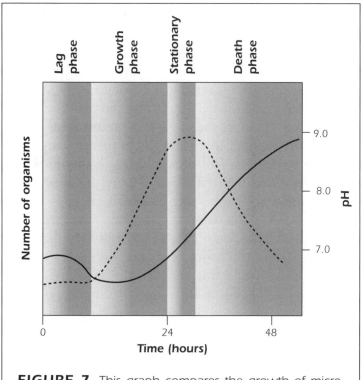

FIGURE 7 This graph compares the growth of micro-organisms (dotted line) and changes in pH (solid line) over time.

are used up. The period in which the cells begin to die or *sporulate*—create a protective outer shell called a spore—is called the death phase.

You can also monitor the condition of your sample by periodically testing its pH. pH measures the acidity or alkalinity of a solution by determining its hydrogen ion concentration. As populations of bacteria grow in liquid *medium*, the pH of the sample changes. Most bacteria show a distinctive pH pattern during their growth cycle. By monitoring a sample's pH, industrial microbiologists can determine whether the fermentation process is proceeding as planned.

The pH of your sample should follow the curve shown in Figure 7. In the initial part of the pH curve, there is a drop in pH (the sample becomes acidic). When the cells grow and divide, an acid product is released by the bacterial cells. As the sugars or amino acids in the medium are used up, there is a rise in the pH curve (the sample becomes more alkaline). This tells the fermentation scientist that the desired product (enzymes, biopesticides, or antibiotics) is being formed.

At the end of the fermentation run, you can do an *assay* to determine the enzyme activity of the bacterial cells manufactured during fermentation. You can also conduct assays while the bioreactor is running. These should be performed at three specific times: at 20–24 hours (growth phase), at 36 hours (stationary phase), and at 48 hours (death phase).

What You Need
Bioreactor
Rubbing alcohol
Fermentation media
Hot-water bath
Thermometer
Drain bacteria
10-cc syringe pipette without needle
Diluted household bleach (1 part bleach to 9 parts water)
pH paper (range pH 4 to 9)
Petri dish or transparent plastic dish smeared with butter or Crisco™

Caution: Perform this activity under adult supervision. Wear safety goggles and rubber gloves to minimize the risk of contamination. Bleach is corrosive; rubbing alcohol is extremely flammable. Always attach an air filter to the inlet and outlet ports. Do not use any culture media that enhances the growth of disease-causing organisms. Do not grow cultures without oxygen.

Sterilize the top portion of the bioreactor by soaking it in rubbing alcohol for about 10 minutes. Remove the apparatus and shake off any excess liquid. Remember not to get the air filters wet. Place the top assembly on the flask. (The flask should have previously been sterilized using the procedure described in the Technical Notes.)

Instructions for preparing fermentation media are provided in the Technical Notes at the back of this book. Heat the media to 30 to 37°C (86 to 98°F). If the media is too hot, you will kill the microorganisms. Working quickly, partially remove the bioreactor's top assembly and pour 500 mL of fermentation media into the flask.

Attach the fish-tank air pump to the air inlet and adjust the airflow so that it bubbles gently into the media. (This will supply oxygen to bacterial cells. It will also create positive pressure inside the apparatus and prevent contamination.) Add 1 teaspoon of drain bacteria mixture to your bioreactor. Secure the top assembly to the flask. Additional instructions for working with bacteria are provided in the Technical Notes at the back of this book.

For the next 48 hours, note any changes that occur to the bacteria. Pay particular attention to your experiment 22 hours after you add the bacteria.*

*You can accelerate the rate of bacterial growth by increasing the amount of air that enters the apparatus. If you adjust airflow, the mixture may begin to foam. To reduce foaming, either decrease the airflow or turn off the fish-tank air pump for a few minutes.

Evaluate the progress of the fermentation run by withdrawing small samples of broth and measuring its pH. Test the samples 22, 36, and 48 hours after starting the run. To remove some of your culture broth, attach the syringe (without needle) to the sampling port. Loosen the screw compress clamp and pull on the syringe handle to withdraw 2 to 3 mL of fermentation broth. Tighten the screw compress clamp and rinse the end of the sampling port with a small amount of bleach solution to prevent contamination of the fermenter. Place a drop of broth culture on the pH paper and record the pH value in your lab notebook. Dispose of the sample as outlined in the Technical Notes (under Working with Bacteria) at the back of the book.

To monitor the enzyme activity of the bacterial cells, withdraw 2- to 3-mL samples from the bioreactor 22 hours, 36 hours, and 48 hours after turning on the bioreactor. Use the procedure outlined above. Spread your sample on a petri dish or small transparent plastic dish smeared with butter or Crisco. Cover your dish and do not disturb it for 24 hours. Then, observe any physical changes that have taken place on the butter or Crisco. Which of your samples (22 hours, 36 hours, or 48 hours) was most effective at breaking up the fatty substance?

Dispose of the sample as outlined in the Technical Notes. Wash your hands thoroughly.

Doing More

- To add a control to this project, add unused fermentation media to a petri dish or transparent dish smeared with butter or Crisco. What changes do you observe after 24 hours?

- Grow other nonpathogenic microorganisms to produce a wide variety of products, ranging from bio-insecticides to antibiotics. Appropriate Bacillus strains are available from science supply companies.

- Investigate the ideal growth conditions for microorganisms by varying the oxygen content, fermentation media composition, starting pH, and temperature in your bioreactor. This experiment will help you understand the process that scientists follow when they are trying to create the most efficient industrial process.

WHAT IS YEAST'S FAVORITE FOOD?

Fermentation is really nothing more than a process used to obtain energy from food. In this process, microbes break down food particles to release stored energy. Do the following project to determine which food sources provide fermenting yeast with the most energy. You will also learn more about the various steps of the fermentation process.

What You Need	
Thermometer	2 Thermos bottles
2-hole rubber stoppers	Permanent marker
Rigid plastic tubing	Masking tape
Vinyl airline tubing	Baker's yeast
Erlenmeyer flask	Limewater
Tap water	Lactose (milk sugar)
Sucrose (table sugar)	

Caution: Do this project under adult supervision. Wear safety goggles, insulated gloves, and an apron or a smock. Limewater is corrosive; do not let it come into contact with your skin or eyes. Keep the container tightly closed to avoid spills.

Insert the thermometer through one hole in one of the rubber stoppers. Insert rigid plastic tubing into the other hole. Attach a 10- to 12-inch piece of vinyl airline tubing to the top of the rigid plastic tubing. Insert two pieces of the rigid plastic tubing into the other rubber stopper and attach the loose end of the vinyl airline tubing from the first rubber stopper to one of them. Place this stopper into the flask. See Figure 8.

Add 1 cup (240 mL) of tap water and 2 teaspoons (10 g) of sucrose to one of the Thermos bottles. Using permanent marker and masking tape, label this Thermos A. Stir the mixture thoroughly. Add a pinch of yeast to about 1 ounce of water. Then, add 1 ounce (30 mL) of the yeast-water mixture to the Thermos bottle. Gently mix the contents. Remove the rubber stopper from the flask and add about $\frac{1}{3}$ cup (80 mL) of limewater. Place the stopper securely on the flask. Place the other stopper on Thermos A.

Record the temperature every hour for several hours and note any changes in the appearance of the limewater. As the yeast converts the sucrose into carbon dioxide and releases energy in the form of heat, the pH of the solution decreases and the limewater forms a precipitate.

Add 1 cup (240 mL) of tap water and 2 teaspoons (10 g) of lactose to the other Thermos bottle. Label this Thermos B. Repeat the procedure outlined above for the contents of Thermos B. Is the temperature during fermentation greater for the sucrose or the lactose? Which

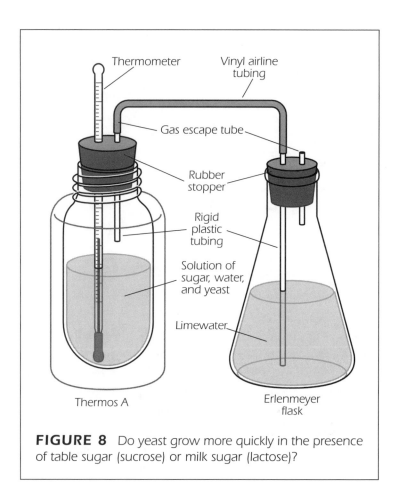

Thermometer

Vinyl airline tubing

Gas escape tube

Rubber stopper

Rigid plastic tubing

Solution of sugar, water, and yeast

Limewater

Thermos A

Erlenmeyer flask

FIGURE 8 Do yeast grow more quickly in the presence of table sugar (sucrose) or milk sugar (lactose)?

sugar causes more precipitate to form? How do you account for these differences? Which of the two sugars is a better source of food for the yeast?

CHAPTER 2

BIOTECHNOLOGY STARTS WITH DNA

Your body contains roughly 100 trillion cells. If you looked inside each cell, you would see *chromosomes* within a nucleus. Chromosomes are made of DNA. No other substance is as important as DNA. It contains a set of chemical instructions that controls all of our cells' activities. When a male's sperm unites with a female's egg, each parent's genetic code is passed on to a new generation.

The search for the source of genetic information began more than a century ago. In 1868, Swiss scientist Friedrich Miescher discovered a method for separating a cell's nucleus from the rest of the cell. He found that cell nuclei contain an acid, which he named "nuclein." This material, which we now call "nucleic acid," is part of DNA.

In the 1950s, scientists used chromatography and X-ray diffraction techniques to build a scale model of DNA. As shown in Figure 9, DNA is made up of subunits called nucleotides linked together in such a way that the result-

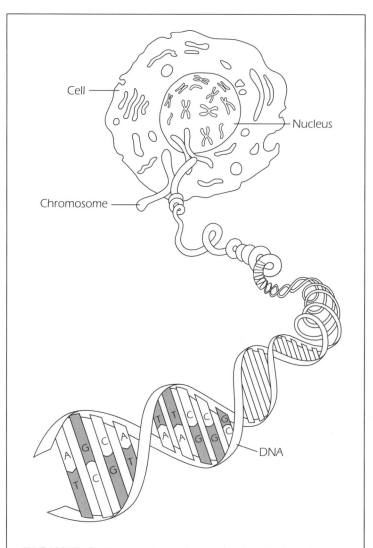

FIGURE 9 The nucleus of every cell contains chromosomes. A chromosome is made of protein and long chains of DNA nucleotide bases. There are four different types of nucleotide bases: guanine (G), cytosine (C), thymine (T), and adenine (A).

ing molecule resembles a long, twisted ladder. Each leg of the ladder consists of alternating sugar and phosphate units. The rungs of the ladder are made of pairs of nucleotide bases that are bonded together.

There are four types of nucleotide bases: adenine (A), thymine (T), cytosine (C) and guanine (G). These bases combine in very specific ways. Adenine pairs only with thymine, and cytosine pairs only with guanine. The order, or *sequence*, of nucleotide bases in a molecule of DNA provides instructions used to build amino acids and link them together into proteins. Proteins control most of the chemical reactions that carry out cellular activities. Without proteins, our bodies could not grow or function. A piece of DNA that contains all the information needed to build one protein is called a *gene*.

A LOOK AT DNA
EXTRACTING DNA

You might think that it's impossible to see DNA without a microscope, but the following project gives you the opportunity to do just that. You will extract DNA and examine its physical characteristics.

Caution: Wear safety goggles and heat-insulating gloves when handling hot objects. Have ice available. Wear safety goggles and rubber gloves when handling rubbing alcohol. Remember, rubbing alcohol is an extremely flammable liquid.

Prepare a detergent/salt solution by adding 10 mL of light-colored liquid dishwashing detergent or shampoo (Dawn™ or Suave™; Daily Clarifying Shampoo™) and 1.5 g of table salt to 100 mL of water. Stir the mixture until the salt is completely dissolved. (Stir gently to avoid foaming.) Remove the skin and first few layers of a large onion.

10 mL light-colored liquid dishwashing detergent or shampoo	Ice bucket with ice
	Strainer
1.5 g table salt	Coffee filter (#6) or cheesecloth
100 mL water	Glass bowl
Large onion	DNA spooler (thin wire loop)
Knife	
240-mL measuring cup	Test tube or clear glass vial
Saucepan for hot-water bath	10 mL 95% ethyl alcohol or rubbing alcohol (stored on ice in a test tube)
Hot tap water (55°C)	
Thermometer	
Spoon	

Coarsely chop the center of the onion and place the onion pieces in a 240-mL (1-cup) measuring cup.

Add enough of the detergent/salt solution to cover the onion pieces. The detergent breaks down the lipids and proteins that make up the cells' outer and nuclear membranes. As a result, DNA is released.

Place the measuring cup in a hot-water bath at 55 to 60°C for about 10 minutes. While the mixture is heating, press the chopped onion against the side of the measuring cup with the back of a spoon. Cool the mixture in an ice bath for 5 minutes. Continue to press the chopped onion against the side of the measuring cup. Rapid cooling will slow the breakdown of DNA.

Using a strainer lined with a #6 coffee filter or four layers of cheesecloth, separate the liquid portion of the onion mixture into a glass bowl. Keep this solution refrigerated or on ice.

Bend the end of a 10-inch piece of thin wire upward into the shape of a fishhook. This will serve as a DNA spooler. Pour 5 mL (1 teaspoon) of your onion extract into a test tube or glass vial. Hold the test tube in one hand at a 45° angle. Slowly add 5 mL of ice-cold alcohol (stored on ice) down one side of the tube, as shown in Figure 10. The alcohol should form a distinct layer over your onion extract mixture. Avoid shaking or mixing the layers.

Let the mixture stand undisturbed for 2 to 3 minutes. The alcohol will cause the DNA to come out of solution. You will see it floating in the alcohol layer.

Lower your DNA spooler into the test tube until the hooked end is at the boundary between the onion extract mixture and alcohol layer. Gently swirl the DNA spooler at the interface of the two layers to wind the DNA strands around the spooler. Use the same motion you would use to wind spaghetti around a fork.

Lift the DNA spooler out of the test tube and carefully observe the extracted DNA. Try to describe its color and physical characteristics in your lab notebook. What does it remind you of? Your extracted DNA will look like a whitish mucus and should have a viscosity and elasticity similar to honey or maple syrup. ("Viscosity" is a fluid's resistance to flow, and "elasticity" is a material's ability to resume its original size after being stretched.)

Place the extracted DNA sample on the lip of the test tube and carefully tug on one end of the mass. When the

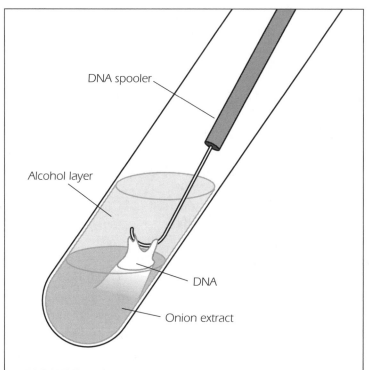

FIGURE 10 The onion's DNA will form a mucus like blob along the interface between the onion extract and the alcohol layer. This material can be removed from the test tube or vial with a DNA spooler.

DNA is stretched, it forms thin, hairlike threads. If you pull the thread too far, it will snap.

Doing More
- Use the same procedure to extract DNA from other sources. Try an animal liver, thymus gland, or even powdered yeast cells.

- Determine whether there physical differences between DNA derived from plant and animal cells. Is one type of DNA more viscous? Can one be stretched farther than the other?

- During the DNA extraction process, you observed some of DNA's physical characteristics such as viscosity and elasticity. Compare the characteristics of other long-chain macromolecules (maple syrup, corn syrup, honey, glycerin, or motor oil) to those of DNA.

THE EFFECT OF HEAT ON EXTRACTED DNA

When DNA is heated to above 165°F (75°C), its viscosity decreases. This change occurs because the molecule is *denatured*. In other words, the bonds between nucleotide bases are broken and the two halves of the DNA ladder fall apart. DNA molecules with a higher ratio of adenine-thymine base pairs tend to denature at a lower temperature than DNA molecules with more guanine-cytosine base pairs. This is because cytosine and guanine are held together by three bonds, while adenine and thymine are held together by just two bonds. See Figure 11.

What You Need	
Extracted DNA sample	Hot plate
Test tube or clear plastic vial	Ring stand
	Ring clamp
Beaker or small pot	Thermometer
Water	

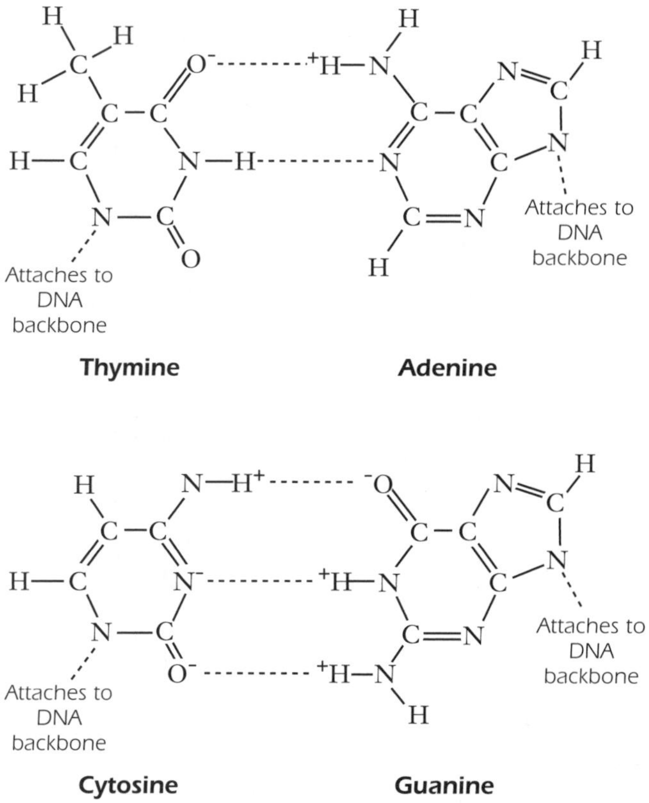

FIGURE 11 The nucleotide base adenine bonds only with thymine, while guanine bonds only with cytosine.

Caution: Be sure to use heat-insulating gloves when handling hot objects. Do this procedure under adult supervision.

Place your extracted DNA sample into a small vial or tube. Fill a beaker or small pot with water and place it on a hot plate. The hot plate should be on a ring stand. Attach

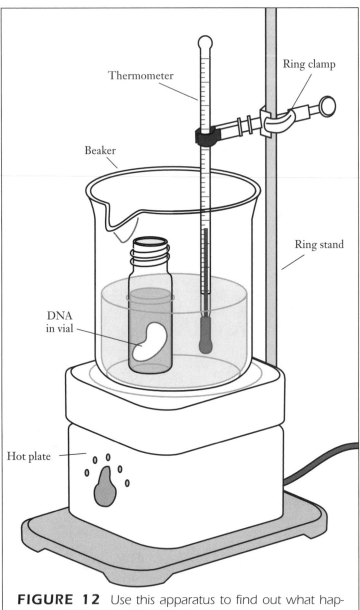

Thermometer

Ring clamp

Beaker

Ring stand

DNA
in vial

Hot plate

FIGURE 12 Use this apparatus to find out what happens to DNA when it is heated.

a hot plate. The hot plate should be on a ring stand. Attach a thermometer to the ring clamp, as shown in Figure 12.

Slowly heat the water bath. Each time the temperature increases about 41°F (5°C), remove the test tube from the water bath and examine the DNA closely. Record changes in your lab notebook.

After each observation, return the test tube to the water bath and continue to heat it until the bath reaches 150°F (65°C). What happens to the viscosity and elasticity of the DNA as its temperature increases?

Doing More

- Compare how extracted DNA from plants (onion) and animals (liver or thymus) react to heat. Is there a difference? If so, how do you account for it?

- Repeat the procedure with protists such as Euglena, Chlorella, or Paramecium. You can grow your own using the procedures described in the Technical Notes. For monerans, use dried Spirulina, which is available at most health-food stores.

ELECTROPHORESING EXTRACTED DNA

Although your extracted samples are not pure, they can still be electrophoresed to confirm that DNA is present. Insert a denatured sample—heated to 194°F (90°C) for 10 minutes—into one well and a nondenatured sample into a separate well. Chapter 1 has instructions for building and using an electrophoresis chamber.

Before beginning this project, please see the What You Need section on page 46.

Place each of your extracted samples in 2 to 3 drops of running buffer and 1 drop of loading dye (probably available from your science teacher). Prepare an 0.4% agarose gel and load 1 drop of each DNA-dye mixture into separate wells.

At the end of your electrophoresis run, stain the gel with a DNA stain using the procedure described in the Technical Notes at the back of this book. Depending on the purity and denaturation of your DNA sample, you may only be able to observe a smear, rather than distinct bands. Observe any differences in migration rates and banding formation between your two DNA samples. How do you account for these differences?

STUDYING FRUIT FLY CHROMOSOMES

The fruit fly, Drosophila, has four distinct life stages: egg, larva, pupa, and adult. See Figure 13. A mating pair of fruit flies produces new adults in just 2 weeks. Larvae hatch 1 day after eggs are laid. During this 6-day stage, a larva molts—sheds its cuticle, mouth, hooks, and spiracles—twice. The growth periods before, between, and after molts are called *instars*. The cuticle (skin covering) of the third instar hardens and darkens to become the

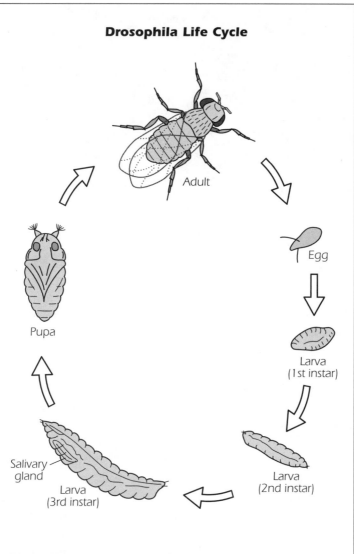

Drosophila Life Cycle

Adult

Egg

Larva
(1st instar)

Larva
(2nd instar)

Larva
(3rd instar)

Salivary
gland

Pupa

FIGURE 13 Drosophila have a relatively short life. A fly reaches adulthood about 2 weeks after hatching and lives as an adult for only a few weeks.

A fruit fly

pupa. The pupa stage is 8 days long. When metamor-
phosis is complete, an adult emerges. Adult flies live for
several weeks. A female can begin laying eggs 2 days after
emerging and remains fertile until it dies.

Fruit flies are often used to study genetics and heredity because they are easy to keep. They are also good for
genetic studies because they have short life cycles, many
obvious hereditary traits, and only four pairs of chromosomes. Fruit fly chromosomes are 100 times larger than
human chromosomes, so they are easy to extract and
observe under a light microscope. With the aid of special
staining techniques, scientists have discovered that fruit
fly chromosomes are composed of a series of alternating
dark and light bands. The dark bands show the location
of genes.

In this project you will have an opportunity to study
the chromosome structure of cells in the salivary gland of
a third-instar Drosophila larva. You will extract the chro-

mosomes from a fly larva and prepare a microscope slide squash to observe under the microscope. Additional information about working with Drosophila is provided in the Technical Notes at the back of this book.

What You Need	
Drosophila larva culture (virilis or mojavensis strain)	Forceps
	Dissecting probe
Microscope slides	Chromosome stain (aceto-orcein)
Medicine dropper	Coverslips
Tap water	Paper towels
Plate	Compound microscope with a 40X objective
Access to a refrigerator	
Dissecting microscope or magnifying lens (10X)	Duco™ cement or colorless nail polish

Caution: Wear safety goggles, rubber gloves, and an apron or a smock when working with stains. Do this procedure under adult supervision.

Place a third-instar larva from the Drosophila culture on a microscope slide with a drop of water. (Third-instar larvae are slow-moving and begin to appear about 6 days after the eggs are laid.) Carefully place the slide on a plate and refrigerate for 20 to 30 minutes to chill and anesthetize the larva.

Following anesthetization, place the slide under a dissecting microscope. If a dissecting microscope is not available, use a magnifying lens. Locate the larva's black mouth hooks. While holding down the larva's abdomen with forceps, use a dissecting probe to pull the larva's head away from its body. You should be able to see the larva's salivary glands. They will look shiny and may still be attached to the head. Carefully tease the salivary glands from the rest of the tissues and place them on a clean microscope slide.

Add 3 to 4 drops of aceto-orcein stain directly to the glands. After several minutes, add a coverslip and remove any excess stain by placing a paper towel at the edge of the coverslip. To crush the sample, gently push down on the coverslip with your thumb or the eraser end of a pencil. This will spread the chromosomes so that they can be easily studied.

View the slide with a compound microscope. First, locate the chromosomes under low power (100X), then examine the banding pattern under high power (400X). Make an accurate drawing of your observations in your lab notebook. Seal the edges of the coverslip with Duco cement or colorless nail polish to preserve the sample.

CHAPTER 3

HEALTHIER PLANTS, TASTIER FOODS

O nce scientists understood the structure of DNA, they started looking for ways to manipulate it in order to manufacture large quantities of particular proteins. For example, they wanted to make insulin to treat *diabetes*.

At that time, insulin was extracted from the pancreatic glands of animals. Because it was difficult, time-consuming, and expensive to obtain and purify insulin, scientists hoped to find a better way of producing it. Scientists also thought that it might be possible to alter the DNA of crop plants to make them resist diseases and insect infestation.

In 1973, biochemists Stanley Cohen and Herbert Boyer developed a technique for altering the DNA of living organisms. They removed a short segment of DNA from the cell of a toad and inserted it into the chromosome of the bacterium *Escherichia coli*. When the bacteria

TABLE 1: Crops Improved by Biotechnology

Plant	Genetic Alterations
Canola	Herbicide tolerance; hybrids
Cantaloupe	Virus protection
Citrus	Salt tolerance
Cotton	Insect protection; herbicide tolerance
Pepper	Virus protection; flavor enhancement
Potato	Increased starch content; insect, virus, and nematode protection; increased cold tolerance
Rice	Virus protection; lower levels of amylose and allergens
Soybean	Herbicide tolerance; improved oil content; fungus protection
Sugar beet	Herbicide tolerance; virus and fungus protection
Tomato	Virus and fungus protection; delayed ripening

reproduced, all of its offspring contained the segment of toad DNA. This technique is now known as *genetic engineering*.

Today, genetic engineering is used to produce a variety of medicines and create crop plants that produce more food and resist pests. Recently, scientists have succeeded in creating cotton plants that fill the hollow middle of their fibers with a small amount of plastic. Soon farmers may be able to grow plastics more inexpensively than chemical manufacturers can synthesize them.

The need for genetic engineering of crops is likely to increase in the future. Experts predict that by 2030, there will be more than 10 billion people living on our planet. We will need three times more food than we produce today to feed them all. At the same time, land that has been intensely cultivated in the past is becoming less and less productive.

Although pesticides and fertilizers have been used to increase crop yields in the past, scientists are concerned about the effect these chemicals are having on the environment. Biotechnology offers solutions to all these problems. Genetic engineering is already being used to alter the physical characteristics of some crops. See Table 1.

THE FIRST GENETIC ENGINEER

The bacterium *Agrobacterium tumefaciens* is an old and experienced genetic engineer. This soil-inhabiting bacterium has been inserting its genes into plant cell chromosomes for thousands of years. When *A. tumefaciens* inserts its genetic material into a plant, the plant produces a tumorous mass—called a gall—at or near the plant's crown, the place where the root and stem meet. The plant cells inside the crown gall tumor manufacture special amino acids—called opines—that serve as food for the bacteria.

Not all plants are susceptible to *A. tumefaciens* and other species of agrobacteria. While they do infect sunflower, grape, apple, rose, peach, willow, and tomato plants, they do not infect the most important crop plants—the grains (wheat, rye, barley, rice).

CROWN GALL TUMORS: A CLOSER LOOK

No matter where you live, you can probably find naturally occurring crown galls on trees and other plants growing in your yard or along your street. You can also look for them in local parks, fields, orchards, or wooded areas. If you have trouble spotting them, ask someone working at a garden center, greenhouse, local county agricultural extension office, or the biology department of a college or university.

When you find a crown gall tumor, cut it off the plant and bring it home. The following project will allow you to see the bacteria growing inside.

What You Need	
Crown gall tumor	Bottle of bacteria stain (1% crystal violet, 1% methylene blue, or 0.25% safranin stain)
Scalpel or sharp knife	
Toothpick	
Microscope slide	Paper towels
Wooden clothespin	Access to a sink
	Compound microscope

Caution: Be careful when cutting the gall. Wear safety goggles, rubber gloves, and an apron or a smock when working with stains.

Carefully cut open the gall and use a toothpick to remove tumor tissue. Spread the tissue on a clean microscope slide. Clip the clothespin to one end of the slide and use it to hold the slide.

Gall tumors, such as this one, are often easy to spot. To find a crown gall tumor, you must look carefully at the plants growing in a field or wood area. Crown gall tumors occur where a plant's roots and stem meet.

Flood the tissue with several drops of stain. Let stand 5 to 10 seconds. Rinse the slide under a gently dripping faucet to remove excess stain. Allow the slide to air-dry on a paper towel. Wash your hands thoroughly.

Observe the slide under a compound microscope at high-dry magnification (430X). Look for rod-shaped specks (agrobacteria) among disorganized masses of plant (gall) tissue.

GROW YOUR OWN AGROBACTERIA

To grow agrobacteria, you will need to *incubate* them. Scientists incubate a sample of cells or organisms in order to promote growth and reproduction. Some organisms—

What You Need	
Transfer chamber	Petri dishes containing nutrient agar
Paper towels	
Crown gall tumor	Bowl
Sterile gauze pad	Microscope slide
Knife	Wooden clothespin
Diluted household bleach (1 part bleach to 9 parts water)	Bottle of bacteria stain (1% crystal violet, 1% methylene blue, or 0.25% safranin stain)
Box of toothpicks with a small hole in one corner	Compound microscope
Access to a sink	Medicine dropper

such as agrobacteria—grow best at room temperature, but many do better in a warmer environment. When scientists use the word "incubation," they usually mean placing a sample of living cells or organisms in an environmentally controlled chamber maintained at a specific temperature for a given period of time. In some cases, environmental factors such as light, humidity, and radiation are also controlled.

Caution: Wear safety goggles, rubber gloves, and an apron or a smock when working with bleach and stains. Do this project under adult supervision.

Follow the procedure outlined in Chapter 1 to build and sterilize a transfer chamber. Put a few clean paper towels inside the chamber. Place the gall, the sterile gauze pad, the knife, the bleach solution, the box of toothpicks, and a petri dish inside the transfer chamber. You may wish to culture several petri dishes. If so, put all the dishes in the transfer chamber now.

Working inside the transfer chamber, place the gall in the bowl and add enough diluted bleach solution to cover it. This will kill any bacteria present on the outside surface. After 10 minutes, lift the gall using the gauze pad. Keep the gall wrapped in this sterile covering. (Do not touch the gall with your fingers.)

Dip the end of the knife blade into the diluted bleach for about 1 minute to sterilize it. Remove the blade and allow it to air-dry. Do not let the blade of the knife touch any surfaces. Pick up the gall and gently unwrap it to expose part of the surface. Carefully puncture the exposed surface with the tip of the knife. Remove one toothpick from the box through the small hole in the corner. Be sure to touch only one end of the toothpick. Do not let the other end of the toothpick touch any surface. Insert the "clean" end of the toothpick into the gall and move it around so that it picks up some of the tissue.

Carefully remove the toothpick with one hand and raise the lid of the petri dish with the other hand. Streak the tip of the toothpick on the agar. Gently place the lid of the petri dish on top of the bottom half of the dish. Wash your hands with soap and water. Incubate the petri dish at room temperature for 3 to 4 days. By this time, bacterial colonies should be growing on the agar.

Place a drop of water on a clean microscope slide. Raise the lid of the petri dish and touch the tip of a clean toothpick to the surface of a bacterial colony. Replace the lid. Immerse the toothpick tip in the water on the slide and swirl it around to transfer the bacteria to the water. Allow the water drop to dry.

Make a stained smear using the procedure outlined in the previous project and view the sample under the microscope. View under high-dry (430X) or oil immersion (1,000X) magnification. Draw what you see in your lab notebook.

Using a medicine dropper, add several drops of diluted bleach to the petri dish before disposing of it. (The bleach should just cover the surface of the petri dish.) Use the bleach to sterilize any objects or surfaces that came into contact with infected plant tissue. Wash your hands thoroughly.

INOCULATING A PLANT WITH AGROBACTERIA

When biologists use the word *inoculation* they mean using sterile technique to deposit material (the inoculum) on or into a sterile growth medium or a living organism. In most cases, the goal of inoculation is to start an organism culture that is subsequently incubated to produce large numbers.

Caution: The sale of agrobacteria is carefully monitored. Ask your teacher to help you obtain a sample. Wear safety goggles, rubber gloves, and an apron or a smock when working with bleach. Do this project under adult supervision.

Carefully tip the box of toothpicks so that a single toothpick can be extracted through the small hole in the box. Be sure to touch only one end of the toothpick. Do not let the other end of the toothpick touch any surface. Use the sterile tip of the toothpick to "wound" two sites on the stem of a sunflower or tomato plant. (One site will serve as a control.)

Follow the procedure outlined in Chapter 1 to build and sterilize a transfer chamber. Put a few clean paper towels inside the chamber. Place the agrobacteria culture and box of toothpicks inside the transfer chamber. Raise the lid of the petri dish and lightly press the tip of a clean toothpick against a bacteria colony. Withdraw the toothpick, and lower the petri dish lid.

Inside the transfer chamber, carefully insert the tip of the toothpick into one of the plant's wound sites. Disinfect the toothpick with diluted bleach and discard it. Using a tag, label this site "Experiment." Insert a clean toothpick into the plant's second wound site. Using a tag, label this site "Control." Wash your hands thoroughly.

Place the plant in a sunny window and water it daily. You should begin to see a gall after about 2 weeks. Continue to watch the gall for another 2 weeks. Periodically draw what you see in your lab notebook. Label each drawing with the date and time. Do you observe any galls forming? Do galls form at both wound sites? Can you reisolate agrobacteria?

MAKING PLANTS INTO FOOD FACTORIES

Although *A. tumefaciens* was first isolated from a crown gall tumor in 1907, it was not until the 1970s that scientists began to understand how the bacterium manipulates a plant's cells to grow food. The genetic material that contains the instructions for converting plant cells into food producers is located on a circular piece of DNA— called a *plasmid*—within each bacterial cell.

When *A. tumefaciens* infects a plant cell, a segment of the tumor-inducing (Ti) plasmid incorporates itself into the plant's DNA. Figure 14 outlines the steps of this process. The following project will show you that the agrobacterium's plasmid is directly involved in tumor formation.

Caution: The sale of agrobacteria is carefully monitored. Ask your science teacher to help you obtain a sample. Wear safety goggles, rubber gloves, and an apron or a smock when working with stains. Do this activity under adult supervision.

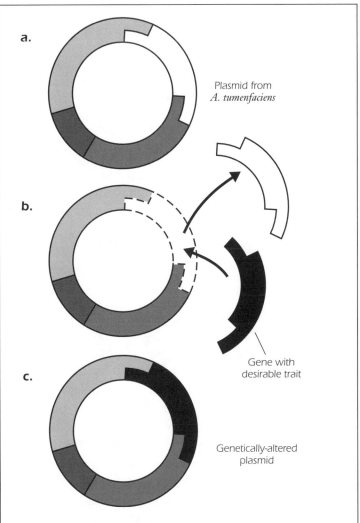

a.

Plasmid from
A. tumenfaciens

b.

Gene with
desirable trait

c.

Genetically-altered
plasmid

FIGURE 14 *A. tumenfaciens* has a special type of circular DNA called a plasmid. Scientists can remove part of the plasmid and replace it with the plant gene they wish to reproduce. When the bacteria is inserted into a healthy plant, that gene becomes incorporated into the plant's DNA.

What You Need	
Two agrobacteria cultures on petri dishes	Diluted household bleach (1 part bleach to 9 parts water)
Incubator (two-egg chick size)	Transfer chamber
Small tags	Box of toothpicks with a small opening in one corner
Permanent marker	
Two petri dishes containing nutrient agar	Potted tomato or sunflower plant

Place one of the two agrobacteria cultures in an incubator for 2 to 3 days. The temperature should be above 37°C (98°F) at all times. Keep the other agrobacteria culture at room temperature (25°C [77°F]) for the same period of time. Both dishes should be kept upside down to prevent condensation. Make sure each dish is marked with its incubation temperature. Sterilize another two petri dishes using the technique described in the Technical Notes at the back of this book. Place all four petri dishes and the box of toothpicks in the transfer chamber.

Carefully remove a toothpick from the box with one hand and raise the lid of the petri dish that was in the incubator with the other. Press the "clean" end of the toothpick to one of the bacterial colonies. Close the lid of the petri dish. Raise the lid of one of the newly sterilized petri dishes and streak the tip of the toothpick on the agar. Replace the lid. Put this dish in the incubator for 3 to 4 days.

Repeating the procedure described above, transfer bacteria from the petri dish at room temperature to the second recently sterilized dish. This sample should be maintained at room temperature for 3 to 4 days.

Inoculate tomato or sunflower plant stems with bacteria grown at each of the two temperatures as described in the previous project. Plan at least three inoculation sites for each temperature. Do any galls form? Do they form at control sites? Did the galls grow on plants infected with agrobacteria grown at the higher temperature?

BIGGER, BETTER, FRESHER, FASTER

Did you know that corn harvested 1,000 years ago had ears the size of your little finger? Farmers have been trying to breed better plants since agriculture began. They have always looked for plants that produce higher yields, are easier to harvest, can be grown in harsh climates, or have fruits that ripen more slowly.

ROSY-RED TOMATOES IN WINTER

If you go to the grocery store in the winter, you may notice that most of the tomatoes look pretty unappetizing. They are red, but they look a little pale. If you bought some, took them home, and ate them for supper, you'd see that they don't taste nearly as good as the tomatoes picked from a local garden in the summer.

These inferior tomatoes were picked before they were ripe, then shipped to various parts of the country, and then sprayed with a chemical called ethylene. Ethylene is a natural plant hormone that causes ripening. The ethylene makes the tomatoes look ripe, but it cannot provide all the sugars that are incorporated into tomatoes when they ripen naturally on the vine. That's why these tomatoes don't taste as good as tomatoes that come from local farms in the summer.

In the last few years, biotechnologists have developed a way of genetically engineering tomato plants to suppress the release of natural ethylene and slow the ripening process. The commercial result is the Endless Summer tomato, which received U. S. Department of Agriculture (USDA) approval in early 1995. These tomatoes do not have to be picked while they are still green, so they taste just as good as tomatoes grown on a local farm—even though they may have been shipped hundreds of miles. You can investigate ethylene production by doing the activities below.

You will need some bananas and six unripened plums, tomatoes, or peaches that are about the same size. Place one piece of fruit in each of the following storage conditions: inside a brown paper bag at room temperature, inside a brown paper bag with a single banana at room temperature, alone at room temperature, alone in a refrigerator, inside a brown bag in a refrigerator, and inside a brown bag with a single banana in a refrigerator. Record how many days it takes for each piece of fruit to ripen. What does the banana do? Does refrigeration hinder or speed up the ripening process?

Doing More

- If the Endless Summer tomato is available in your area, repeat the previous activity, using it as a test fruit.

- Look for other commercially available genetically engineered plants. Examples include U.S. Calgene's Flvr Savr tomato, BXN cotton (controls weeds), or

Calgene's canola oil (high concentration of lauric acid makes it superior for salads and cooking).

- Research how transgenic livestock are being used as drug factories to produce therapeutic proteins in their milk.

FLOWERS IN A HURRY

It is a few days before Christmas. Business is booming for florists in Pasadena, California. Orders are pouring in for roses, orchids, and other flowers to be used on floats at the Tournament of Roses parade that precedes the Rose Bowl football game on New Year's Day.

A few decades ago, this might have been a problem. Today, plant growers can use a method called micro-propagation to grow flowering plants in a hurry. First, plants are treated with a hormone that stimulates budding. Once buds begin to grow, they are cut from the plant and treated with a culture medium to promote roots. Thousands of identical plantlets can be produced in a short time using this method.

You can see how micropropagation works by growing *Kalanchoë diagremontiana* plantlets. Tiny plants develop in the notches of the plant's fleshy leaves. Once the plantlets are fully developed, they drop to the ground and grow into mature plants. Grow these plants in a brightly lit area at room temperature.

Compare the number of plantlets produced by plants that produce and drop plantlets naturally to the number produced by a standard propagation technique—taking cuttings from "mother" plants (such as a geranium), placing them in water to encourage root growth, and then planting them in soil. How good a micropropagationist are you?

WAGING WAR
AGAINST WEEDS

Some plants produce chemicals that prevent the growth of other plants in the same geographic area. These *allelopathic* agents act as natural *herbicides*. Juglone is a natural herbicide found in many plants in the walnut family. Investigate the characteristics of juglone by doing the following activity.

Caution: Wear plastic gloves when handling walnut fruit flesh; it will stain your hands (or clothing) brown! Do these experiments under adult supervision.

Collect (or purchase) fresh, raw walnuts from a black walnut tree, preferably in the late summer or fall. Soak the flesh from a crushed walnut in a $\frac{1}{2}$ cup of water overnight. Use this juglone-laced water as the "treatment" to investigate the effect of this water versus rainwater on the growth of young plants (tomato, yard weeds, etc.). Repeat this experiment using two or three crushed walnuts to $\frac{1}{2}$ cup of water. This will give you a higher concentration of juglone. Some scientists believe that walnut roots also produce an allelopathic effect. Design an experiment to test this idea. What about walnut bark?

Remove about 3 inches of topsoil from beneath an actively growing walnut tree. Plant a few young tomato plants (or other plants) in this topsoil. What happens to the plants? Store some of the topsoil for an extended period of time and then plant a few small plants in it. Do the juglones degrade over time? If so, do you think this is a commercial advantage to a manufacturer?

Test some of the following plants to see if they produce natural herbicides: hardy chrysanthemum, spiderwort, crocus, Canadian hemlock, Virginia creeper, pot marigold, dandelion.

PLANTS VERSUS INSECTS

Many plants produce natural agents that are poisonous to insects in the same geographic area. You can learn more about these chemicals in the following activities.

Caution: Do not breathe in the dust released when flowers are ground. It is harmful.

Pyrethrin is a natural insecticide produced by chrysanthemum flowers. You can obtain it by grinding dried chrysanthemum flowers into a fine powder using a mortar and pestle. Place the flower powder in a small pie tin inside a plastic bag. Collect various insects and place them inside the bag. Record what happens to each insect in your lab notebook. Repeat the experiment using talcum powder instead of ground chrysanthemum flowers. Do you get different results?

Visit a local garden center and look for commercial products containing chemically extracted pyrethrins. Put a small amount onto a piece of kitchen sponge and place it in a plastic bag. Collect various insects and put them inside the bag. Record what happens to each insect in your lab notebook. How does this material compare to the ground chrysanthemum flowers?

BACTERIA VERSUS INSECTS

There are more than 90 species of naturally occurring bacteria that target and kill insects. *Bacillus thuringiensis* (Bt) is one of the safest insecticides sold commercially. It produces a substance that causes intestinal paralysis in the larvae of mosquitoes and some moths and butterflies. In recent years, scientists have inserted Bt genes into other microbes to produce more potent "biobullets."

Visit a local garden center and scan the shelves for products that contain Bt. Design some experimental trials that test the products' effectiveness on caterpillars. What is the lowest dose that can kill particular species of caterpillars? Are mosquito larvae more or less sensitive than caterpillars to Bt?

PLANT CLONING

All cells are said to be *totipotent*. This means that they have the ability to develop into a complete organism. This biological principle is the basis for the plant tissue culture—or plant-cloning—industry. A *clone* is a genetically identical copy of an organism. It is possible to control the biological characteristics of a clone. For example, scientists can create an organism that has specific growth requirements, is protected against certain viruses, or grows and reproduces more quickly. In the following project, you will clone a carrot.

PROCEDURE I

OBTAINING STERILE PLANT TISSUE
(5 to 7 days)

Caution: Wash your hands with antibacterial soap before beginning and upon completing this project.
Build and sterilize a transfer chamber using the instructions provided in Chapter 1. Put a few clean paper towels inside the chamber along with the following items: a stack of 3 to 5 paper towels moistened with bleach solution, two unopened sterile petri dishes, a bottle of sterile water, a packet of carrot seeds, 1 ounce of bleach solution,

What You Need	
Transfer chamber	2 baby bottles with covered nipple
Paper towels	
Diluted household bleach (1 part bleach to 9 parts water)	Carrot seeds
	Medicine dropper
	15 alcohol wipes
4 petri dishes	Sterilized forceps
Sterile water (boiled water in capped baby bottles)	A package of 2 × 2-inch sterile gauze pads
	Small container

a medicine dropper, fifteen alcohol wipes, the forceps, a gauze packet, and a small container for waste liquids.

Working inside the transfer chamber, remove the lid of one petri dish and place the top and bottom side by side. Do not touch the interior of the petri dish. Open an alcohol wipe and use it to cleanse the bottom half of the forceps. Open a second alcohol wipe and clean the forceps again. Rest the tips of the forceps inside the lid of the petri dish. Place the bottom half of the petri dish on top of the moistened paper towels. Fill the medicine dropper with bleach solution and squeeze it into the bottom half of the petri dish five times.

Sprinkle 10 to 15 carrot seeds into the petri dish with bleach solution. Gently swirl the seeds and solution once every 2 to 3 minutes for the next 20 minutes to disinfect the seeds' outercoats. Using the forceps, remove any floating seeds. Resterilize the tips of the forceps. Carefully pour the bleach solution into the waste container.

Remove the cap from the baby bottle containing sterile distilled water and rinse the seeds. The procedure for preparing the sterile water bottle is described in the Technical Notes at the back of this book. Pour the water into the waste container. Repeat this step three times.

Open the package of gauze pads without letting the pads touch any surface. Place the second petri dish next to the open dish containing the washed seeds and the opened package of gauze pads. Pick up the gauze with the sterile forceps. Carefully lift the lid off the second petri dish and place the gauze in the center of the bottom piece. (Remember to open the petri dish cover only wide enough to allow for the transfer of the gauze pad.)

Moisten—but do not drench—the gauze pad using sterile distilled water from the baby bottle. Resterilize the forceps and carefully transfer one disinfected carrot seed from the open petri dish to the petri dish containing the moistened gauze pad. (Remember to open the petri dish cover only wide enough to allow for the transfer of the seed.) Repeat this step five times. Be sure to resterilize the forceps between each transfer.

Place the second petri dish in a cool 20 to 25°C, poorly-lit area for 5 to 7 days. Light is not necessary for carrot seeds to germinate.

PROCEDURE II

INITIATION OF STERILE PLANT TISSUE
(4 to 8 weeks)

After about a week of growth, the carrot plants should be large enough to "initiate." This part of the plant-cloning procedure involves wounding the plant to encourage the formation of a callus, a mass of soft tissue that occurs at the site of an injury.

The tissue within the callus is undifferentiated. This means that it can easily be manipulated to grow into a complete new plant. It is possible to induce callus tissue to differentiate in a desired way by adjusting the ratios of auxin and cytokinin, two plant hormones, in a culture medium.

What You Need	
Transfer chamber	6 sterile disposable scalpels (with triangular blades)
Paper towels	
Diluted household bleach (1 part bleach to 9 parts water)	10 alcohol wipes
	Carrot seedlings from Procedure I
2 bottles of callus-inducing media	Ruler

Caution: Wash your hands with antibacterial soap before beginning and upon completing this procedure.

Sterilize your transfer chamber by wiping down the inside surface of the box with a paper towel and diluted bleach. Put a few clean paper towels inside the chamber. Place the following materials inside the transfer chamber: a stack of 3 to 5 paper towels moistened with bleach solution, two bottles of callus-inducing media, six scalpels (in sterile wrappers), and ten alcohol wipes.

Working inside the transfer chamber, place the petri dish containing sterile germinated carrot seeds and the bottles of callus-inducing media on top of the moistened paper towels. Use an alcohol wipe to sterilize the surface of the cap and neck of one bottle containing callus-inducing media. Carefully remove the cap from the bottle and place it on the moist paper towels so that the cap's interior is facing you. Do not touch the interior of the cap or the top of the open bottle to any surface.

Carefully lift the lid of the petri dish containing the germinated carrot seeds. (Open it as little as possible.) Using a scalpel, cut the root or shoot of a carrot seedling into 0.5-cm pieces. Use the point of the scalpel to pick up and transfer one seedling slice onto the surface of the callus-inducing media. It is important to proceed slowly. Do not contaminate any sterile surfaces.

Use a new scalpel to transfer each seedling slice to the bottle containing callus-inducing media. Place no more than three slices into a single bottle. Carefully replace the cap of the media bottle.

Add seedling slices to the second bottle of callus-inducing media by repeating the steps described above. Incubate the bottles horizontally at 23 to 25°C for 4 to 8 weeks. Avoid direct sunlight. Do not remove the caps of the media bottles; make all observations through the glass.

PROCEDURE III

DIFFERENTIATING CELLS INTO PLANTLETS
(4 to 6 weeks)

Within a month, callus tissue should have formed and increased in size about fivefold. If you find that your callus tissue is less than 1 cm in diameter, let it grow for a few more weeks before performing this part of the experiment. If the callus tissue has become contaminated, it should not be used. Discard any contaminated tissue immediately.

What You Need	
Transfer chamber	6 sterile disposable scalpels (with triangular blades)
Paper towels	
Diluted household bleach (1 part bleach to 9 parts water)	2 bottles of differentiation media
Bottle containing callus tissue	10 alcohol wipes
	Growlux fluorescent light bulb

Caution: Be sure to wash your hands with antibacterial soap before beginning and upon completing this procedure.

Sterilize your transfer chamber by wiping down the inside surface of the box with a paper towel and diluted bleach. Put a few clean paper towels inside the chamber. Place the following materials inside the transfer chamber: the bottle containing callus tissue, sterile scalpels (in sterile wrappers), the bottles of differentiation media, and ten alcohol wipes.

Working inside the transfer chamber, use alcohol wipes to sterilize the surface of bottle caps and necks. Carefully remove the caps of the bottles. Place the caps on the paper towels so that each cap's interior is facing you. Do not touch the interior of a cap or the top of an open bottle to any surface.

Carefully unwrap one scalpel and use its point to transfer a piece of callus that is about 1 cm in diameter into one bottle of differentiation media. Use new scalpels to transfer other callus tissue samples into the bottle of differentiation media. Place calluses at least 2 to 3 cm from each other. Replace the bottle cap.

Transfer callus tissue samples to the second bottle of differentiation media by repeating the steps outlined in the previous paragraph. Incubate the bottles of differentiation media horizontally at 23 to 25°C for 4 to 6 weeks. Expose the bottle to a Growlux fluorescent light bulb for 16 hours each day (to simulate daylight). Avoid placing cultures in direct sunlight.

The resulting plantlets were produced entirely from tissue culture. These tender, young plants will have to go through a period of adjustment before they can be successfully transplanted into soil. You can condition the plantlets by exposing them to much higher light-intensity levels.

THE PERFECT LOAF

People have been making bread for thousands of years. Originally, all bread was unleavened, or flat. This type of bread was made by mixing grain meal with water and baking the dough on heated rocks. Today, white bread is made by combining wheat flour, water, salt, and active yeast cultures. Yeast is what makes dough rise. Life just wouldn't be the same without it!

Amylase enzymes in the moistened flour convert starch to glucose, which nourishes the yeast cells. Gluten, the protein that gives bread dough its elasticity, provides the yeast with nitrogen. It also traps the carbon dioxide generated by yeast, causing the dough to rise.

1 g dried yeast	75 g wheat flour
Bowl	Two 12-inch-long clear plastic or glass cylinders
50 mL water	
Wooden spoons with long handles	Ruler
	Stopwatch

The following projects investigate the properties of bread dough.

Place 1 gram of viable dried yeast in a bowl, and add 50 mL tap water. Use a wooden spoon to mix and rehydrate the yeast. Add 75 g of flour to the bowl and mix. Roll the dough into a sausage shape and place it inside one of the cylinders.

Use a ruler to measure the height of the dough and record the value in your lab notebook. Continue to measure the height of the dough at 10-minute intervals for 1 hour. Record these results in your lab notebook, too.

Doing More

- Repeat the experiment using various types of yeast. Which makes the dough rise the fastest? The slowest?

- Repeat the experiment with various types of flour. Does the type of flour affect the rate at which the dough rises? If so, are the differences related to the protein content of the flours?*

* Wheat flour is 12 percent protein; bread flour is 11 percent protein; cake flour is 8.5 percent protein; all-purpose flour is 10.5 percent protein.

- Obtain the enzyme alpha-amylase from a wine-making store or catalog. This enzyme converts starch into glucose. Investigate its effect on the rate at which dough rises.

- Salt interferes with the activity of an enzyme called protease. This enzyme can inhibit gluten, causing dough to become a sticky mass that will not rise. However, too much salt can make bread less elastic (tougher). Try varying the amount of salt in bread dough. What is the optimum amount of salt? Look at packaged bread to see how much salt various manufacturers add to their bread.

BIOTECH DAIRY PRODUCTS
NO NEED FOR REFRIGERATORS

Milk is a key part of the diets of people living in Europe and North America. It has many important nutrients—proteins, sugars, fats, minerals, and vitamins. Unfortunately, it spoils rapidly at temperatures above 50°F. Even though it is transported in refrigerated trucks, it is often difficult keep milk cold.

Scientists have found that kappa casein, a protein in milk, controls the temperature at which milk spoils. A number of newly developed milk products have been genetically engineered to contain very high concentrations of this important protein. These products can be stored at room temperature for extended periods.

See if you can simulate this result by adding a small amount of kappa casein to about 1 ounce of whole milk. Use a moistened toothpick to collect the kappa casein and mix it into the milk. While regular milk spoils in less than

2 days at room temperature, milk treated with kappa casein should last considerably longer—perhaps even more than a week.

MAKING CHEESE

People first started making cheese because it keeps better than milk and has the same nutritional value. Today, more than 400 different types of cheese are sold worldwide.

Cheese-making is a complex process that involves skill and patience. When bacteria and rennet—the stomach lining of an animal—are added to milk, the milk forms curds, which are solid lumps, and whey, a watery liquid. The curds are separated from the whey, salted, and aged in a warm room for several months.

The enzyme chymosin, which is the active ingredient

A worker adds a bowl of rennet to a vat of milk during the cheese-making process.

of rennet, breaks specific bonds in a protein molecule on the surface of milk particles. This reaction forms a new protein called calcium paracaseinate. When this protein combines with calcium ions, it forms a solid curd. Milk fat and water are also incorporated into curds.

Until recently, chymosin was obtained from the stomach of milk-fed calves that used it to digest their mother's milk. In 1990, the Food and Drug Administration approved the use of chymosin produced by genetically engineered bacteria. The genetically engineered chymosin is exactly the same as chymosin produced by calves. To make it, scientists extract DNA with instructions for producing chymosin molecules from calf stomach cells and insert it into bacteria. As the bacteria grow, they produce large amounts of chymosin.

In the following project, you can use genetically engineered chymosin (available at most health-food stores) to make some cheese.

What You Need	
Milk at room temperature	Medicine dropper
Glass measuring cup	Oven or hot-water bath
Chymosin	Kitchen thermometer

Caution: Hot liquids can scald.
Pour about 50 mL (1.5 oz.) of the milk into the glass measuring cup and add a single drop of chymosin.

Carefully swirl the milk gently in the cup for about 10 seconds, then place the cup in the oven or hot-water bath.

[If you are using a hot-water bath, the temperature should be maintained between 95°F (35°C) and 110°F (43°C).] Make sure that the temperature of the milk does not exceed 98°F (37°C). Within 30 minutes, the milk fat and protein (calcium paracaseinate) will form curds.

Doing More

- To compare the potency of genetically engineered chymosin with that of traditional rennet (available in tablet form at health-food stores), crush a rennet tablet between two spoons and then mix the powder with a second 50-mL sample of warm milk. Does curdling take more or less time to occur?

- If possible, weigh the curds to quantify your results. How many grams of curds are produced by a specific quantity of chymosin or rennet? Is using this ratio a good way to evaluate curdling?

- To see if the milk's fat content affects curdling, try the same experiment with 1%, 2%, skim, and whole milk. You may also want to test powdered milk, condensed milk, and goat milk. Try both rennet tablets and bacterial chymosin.

- Investigate the effect of temperature on curdling. Does cold milk curdle faster than very warm milk?

- What temperature range works best to promote curd formation? Do exceptionally hot or cold temperatures inhibit enzyme action?

- Add crushed tablets of dietary calcium (available in health-food stores) to 1.5 oz. of milk. Does curdling occur at a faster rate?

- Search for other other substances that curdle milk. Try adding small amounts of acids (lemon juice or vinegar) or bases (a pinch of sodium bicarbonate) to different samples of milk.

JUMP-STARTING THE NITROGEN CYCLE

You probably know that water is constantly cycling through the environment. It falls to Earth as rain and soaks into the ground. Plants absorb the water through their roots. The water that plants do not use evaporates through their leaves and returns to the atmosphere. Some water also evaporates from the surface of lakes and rivers. But water is not the only material that cycles through ecosystems.

Nitrogen, the most common gas in the air, is an important component of all living things. Plants and animals need nitrogen to make proteins and nucleic acids. But most living things cannot use the nitrogen gas in the air. Nitrogen gas is composed of two atoms that are held together by a very strong bond.

A few types of bacteria can "fix," or split, nitrogen molecules. While some of these bacteria live in the soil, others actually live inside the roots of *legumes*—a group of plants that includes peas, soybeans, alfalfa, and clover. The bacteria give the legumes a constant supply of nitrogen. Nitrogen-fixing bacteria convert nitrogen into ammonia or nitrates, which can be absorbed by plants. Plants use this nitrogen to make the proteins that allow them to grow and function.

When animals eat these plants, they use the plant proteins to build the proteins their bodies need to make hair, skin, and muscles. When a plant or an animal dies, bac-

These root nodules from a bean plant contain
Rhizobium bacteria.

terial decomposers break down their bodies and nitrogen
is released into the soil. Some of this nitrogen forms
nitrogen gas and returns to the atmosphere. Then, the
cycle begins again. See Figure 15 on the next page.

Because plants grow better when they get plenty of
nitrogen, some farmers add nitrogen-rich fertilizers to
their fields. Other farmers would rather not add chemicals to their fields, so they rotate their crops instead. First
they plant corn or wheat. When these crops are harvested, the farmers plant legumes in order to add more nitrogen to the soil. Legumes have knotlike *root nodules* that
contain their nitrogen-fixing bacteria.

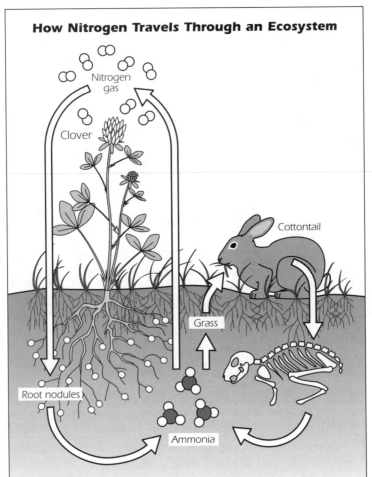

How Nitrogen Travels Through an Ecosystem

Nitrogen gas

Clover

Cottontail

Grass

Root nodules

Ammonia

FIGURE 15 Plants need nitrogen to grow, but they cannot break down nitrogen gas in the atmosphere. Bacteria in the roots of legumes can break apart nitrogen gas, forming ammonia. All plants can use the nitrogen in ammonia to make proteins. When a cottontail eats grass, it uses the plant proteins to build animal proteins. When bacteria break down dead organisms, proteins and other compounds are converted to ammonia. Ammonia that is not absorbed by plants may be converted to nitrogen gas.

You can learn more about nitrogen-fixing bacteria and legumes by investigating clover or crown vetch, which are often seen along the roadsides and in fields. Carefully dig up the root system of a healthy plant and wash off the soil. Do you notice any grain-size nodules on the roots? What's going on inside these nodules? Do the following project to find out.

ROOT NODULES: A CLOSER LOOK

What You Need	
Scalpel or sharp knife	Access to a sink
Root nodule from a legume	Paper towels
2 microscope slides	Compound microscope
Wooden clothespin	
Bottle of bacteria stain (1% crystal violet, 1% methylene blue, or 0.25% safranin stain)	

Caution: Be careful when cutting the root nodule. Wear safety goggles, rubber gloves, and an apron or a smock when working with stains.

Using the scalpel or knife, cut the root nodule into thin slices. Crush one slice between two microscope slides and transfer all of the crushed material to one of the slides. Clip the clothespin to one end of the slide and use it to hold the slide.

Flood the crushed material with several drops of stain and let stand 5 to 10 seconds. Rinse the slide under a gently dripping faucet to remove all excess stain. While the slide is air-drying on a paper towel, wash your hands thoroughly.

Observe the slide under a compound microscope at high-dry magnification (430X). Look for pockets of rod-shaped bacteria in groups of four to eight cells inside a thin membrane. The scientific name of the bacteria you see is *Rhizobium trifolium*.

GROWING YOUR OWN ROOT NODULES

What You Need	
Pea, bean, or lima bean seeds	Pot of boiling water
	Mortar and pestle
Access to a sink	Sterile potting soil
Paper towels	Bottle of bacteria stain
2 small containers	(1% crystal violet, 1% methylene blue, or 0.25% safranin stain)
Water	
Marking pen	Compound microscope
Clover or vetch nodules	

Germinate garden pea, sweet pea, bean, or lima bean seeds on damp paper towels. When they have developed a root system that is 1 to 2 inches (2 to 5 cm) long, divide them into two groups and place each group in a separate container of shallow water. Label one container "Boiled" and the other container "Nonboiled."

Collect 15 to 20 clover or vetch nodules from a nearby field or park. Boil half of the nodules for 15 minutes to kill any nitrogen-fixing bacteria. Grind the boiled nodules using a mortar and pestle. Mix the nodule grindings with the seedlings in the container labeled "Boiled." Next, grind the remaining nodules and add them to the container of seedlings labeled "Nonboiled." The seedlings exposed to the nonboiled nodules will be inoculated with nitrifying bacteria. The seedlings exposed to the boiled nodules will serve as a control.

After a few days, plant all the seedlings in sterile potting soil (obtained at a garden center). After 4 to 8 weeks, dig up the plants and look for root nodules. Stain the nodules and examine them under a microscope. Do all of the plants contain bacteria?

CHAPTER 4

STUDYING
HUMAN
GENES

Genes manufacture the proteins that allow our bodies to develop properly and function efficiently. They regulate how we process nutrients, respond to environmental changes, and fight infections. But when a gene's code has an error, even a very small one, it may produce defective proteins. The result can be physical deformity, disease, or even death.

A *mutation* in genetic material is the underlying cause of more than 4,000 known diseases. These include muscular dystrophy, hemophilia, cystic fibrosis, and Huntington's disease. Even common disorders, such as heart disease and most cancers, are influenced by hereditary factors.

During the past decade, researchers have made great progress toward understanding how genes work and how they are linked to disease. Today, medical scientists can identify genetic mutations that may cause a person to develop a particular disease. This ability has raised the hope of some and the concern of others.

The possibility of overcoming a number of genetic disorders through prevention, early detection, or even direct intervention is one of the most exciting prospects of our time. The flip side of this promising new technology is genetic discrimination. If insurance companies have information about a person's medical future, they may refuse to insure the individual. If a potential employer knows that a person may develop a disease in the next few years, the company may choose to hire someone else. Many people want Congress to pass laws specifying who should—and who should not—be allowed access to a person's genetic profile.

HOW GENES ARE INHERITED

Heredity is the passing of traits or characteristics from one generation to the next. When an egg and a sperm unite, a copy of each parent's genes is incorporated into the resulting embryo. Humans have about 100,000 genes spread over 46 chromosomes. Half of these chromosomes come from the person's mother, and the other half come from the person's father. Thus, we have 23 pairs of chromosomes in the nucleus of every cell.

Each pair of chromosomes has very similar genetic material. The *laws of dominance* determine whether a child will display the trait coded for by a gene inherited from its mother or its father. Some genes, called dominant genes, express themselves when only one copy is present. Other genes, called recessive genes, express themselves only when two copies are present. Scientists use capital letters to designate dominant genes and lowercase letters to represent recessive ones. An individual with two recessive genes for a trait is said to be *homozygous* for that trait. An individual with one dominant gene and one recessive gene is said to be *heterozygous*.

The genetic makeup of an individual is known as a genotype, while the observable physical characteristic that results from an individual's genotype is known as a phenotype. In humans, gender is determined by two sex chromosomes. Individuals with two X chromosomes (XX) are females, whereas those with one X chromosome and one Y chromosome (XY) are males.

A LOOK AT TRAITS

Can you roll your tongue? Do you have a widow's peak? Do you have attached earlobes? All these characteristics are inherited. By studying large groups of people, scientists have determined which of these traits are dominant and which are recessive.

Use the information in Table 2 to find out how many dominant and recessive traits you have. Compare your results to those of family members, friends, and classmates. Count how many people have each trait and how many do not. Then, calculate the percentage of individuals who exhibit a particular trait. Here's how to calculate the percentage of your friends who are tongue rollers:

$$\frac{\text{Number of Tongue Rollers}}{\text{Total Number}} \times 100 = \text{Percentage of Tongue Rollers}$$

Based on your percent calculations, which traits do most people display?

How Some Characteristics Are Inherited

Trait	Dominant Gene	Recessive Gene
Tongue rolling	Roller	Nonroller
Widow's peak	Present	Absent
Shape of earlobe	Free earlobe	Attached earlobe
Cleft in chin	Absent	Present
Freckles	Present	Absent
Dimples	Present	Absent
Shape of eyebrows	Bushy	Fine
Mid-digital hair	Present	Absent

MAKING A FAMILY PEDIGREE

A pedigree is a diagram showing the transmission of a genetic trait through several generations of a family. It provides clues to the kinds of traits that may be passed on to future offspring. A pedigree is also useful for determining the genotype of an individual, as well as predicting the genotypes of future generations. An example of a pedigree is shown in Figure 16 on the next page.

In a pedigree, each individual is represented by an Arabic number, while each generation is represented by a Roman numeral. Squares represent males and circles represent females. A shaded box indicates an individual who displays the characteristic being studied.

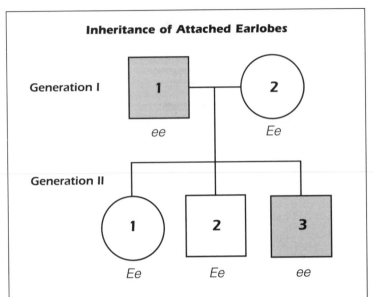

Inheritance of Attached Earlobes

Generation I

1   2

ee   *Ee*

Generation II

1   2   3

Ee   *Ee*   *ee*

FIGURE 16 A pedigree can sometimes be used to determine how a particular genetic trait, such as attached earlobes, is inherited.

Figure 16 looks at how attached earlobes are inherited. The gene responsible for free earlobes is dominant over the gene for attached earlobes. Because a capital letter is used to denote a dominant trait and a lowercase letter is used for recessive traits, we will use (*E*) to represent free earlobes and (*e)* for attached earlobes. Family members with a recessive trait are homozygous for the trait and are represented as (*ee*).

Individual 1 from generation I and individual 3 from generation II have attached earlobes and must be homozygous recessive (*ee*). The other individuals have free earlobes. They could have a genotype of either (*EE*) or (*Ee*). However, since individual 2 must have passed one recessive gene to individual 3, we know that all three individuals with free earlobes must have the genotype (*Ee*).

Trace the pattern of inheritance of a number of traits by constructing a family pedigree. Include your parents, brothers and sisters, grandparents, aunts and uncles, cousins, and nieces and nephews. Number the generations and each family member on your pedigree. Shade in the symbols for those individuals who have the trait. Determine each person's genotype and write it below each symbol. If a family member's genotype is inconclusive, simply write the dominant gene.

USING A PUNNETT SQUARE

Biologists and geneticists use a diagram called a Punnett square to predict all the ways genes can combine during fertilization. Figure 17 shows a Punnett square for the

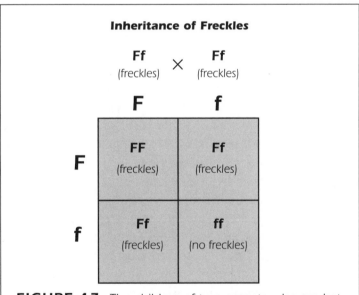

FIGURE 17 The children of two parents who are heterozygous for freckles have a 75 percent chance of having freckles.

inheritance of freckles, which is a dominant trait, when both parents are heterozygous. The parents' genotypes are placed on the outside of the square. Letters inside show the possible gene combinations for an offspring. Is it possible for these parents to have a child with no freckles?

Using information provided by your pedigrees, construct a Punnett square to determine the possible genotype combinations of another child born to your parents or to an aunt and uncle.

DIAGNOSING GENETIC DISEASES

Diseases that run in families are called genetic diseases. What is the risk of inheriting a genetic disease? Why do some diseases appear more often in males than in females? Scientists use family histories, as well as images of chromosomes and molecular studies of DNA, to answer these questions.

Until recently, doctors could not tell whether someone had a genetic disease until symptoms appeared. Gene-screening techniques have made it possible to determine whether a person is predisposed to a certain disease. These tests can also confirm the presence of a specific gene defect or mutation in an individual or a family.

Genetic screening involves examining a person's DNA in order to detect an abnormality that signals a disease or disorder. The defect may be extremely small. In some cases, a genetic disease is caused by a change in, or deletion of, one nucleotide base. Other genetic disorders are caused by large abnormalities. Sometimes segments of DNA detach from one chromosome and attach to another. Other conditions occur when large segments of DNA are missing altogether.

Predictive gene tests identify people at risk for a disease before any symptoms appear. These tests identify disorders that run in families because the defective gene is passed from one generation to the next. More than two dozen of these tests are currently available for diseases such as Tay-Sachs, cystic fibrosis, and Huntington's disease. Scientists are in the process of developing similar tests for Lou Gehrig's disease, some forms of Alzheimer's, extremely high cholesterol levels, and some cancers, such as colon and breast.

To develop predictive gene tests, scientists study DNA samples from members of families with a high incidence (over several generations) of a particular condition. If the gene itself cannot be studied, they look for easily identifiable segments of DNA, known as *genetic markers*, that are consistently inherited by family members with the disease but are not found in relatives who are disease-free.

An accurate gene test can tell whether an individual has the mutation associated with a particular disease, but it cannot determine whether the person will actually develop the disease. That's because many factors influence the gene's expression. Predictive gene tests deal in probabilities, not certainties. In some cases, people who know they are predisposed to a particular disease can modify their diet, behavior, or lifestyle to decrease their risk of developing the disease. In other cases, they can undergo regular screening for the disease. Many cancers can be successfully treated if they are caught early enough.

TECHNIQUES FOR STUDYING CHROMOSOMES

The DNA contained in a single chromosome is a long, highly compact chain made up of 50 million to 250 million

This cytogeneticist is examining a karyotype.

base pairs. Each chromosome contains several thousand genes. To identify defective, extra, or missing chromosomes, scientists prepare a *karyotype*—a visual representation of an individual's chromosomes arranged in a specific way. This standard arrangement shows *homologous pairs* ordered according to their size, shape, banding pattern, and *centromere* position. Karyotypes have become increasingly important as more diseases are linked to chromosomal abnormalities.

Scientists who prepare and study karyotypes are called cytogeneticists. First, they extract chromosomes from white blood cells and expose them to a variety of chemicals and stains to make the chromosomes easier to see. Next, they photograph the cells in the metaphase stage of mitosis (division) through a microscope. The characteristics of chromosomes are easiest to see during metaphase. The cytogeneticists then use computer equipment to arrange the *chromosome spread* into a karyotype. The following project will show you what cytogeneticists do every day.

What You Need	
A photocopier	Copymaster 1
Scissors	Copymaster 2
Tape or glue	Copymaster 3

Make a photocopy of the chromosome spread shown in the photograph on the next page. You may want to use the enlargement option on your photocopier, so that the details of each chromosome are more apparent.

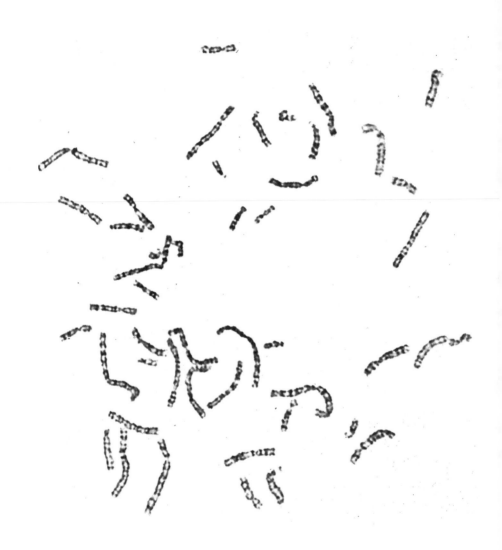

Chromosome spread

Carefully cut out the chromosomes with scissors and arrange them in pairs based on their physical characteristics. Table 3 will help you organize the chromosome spread into a karyotype.

TABLE 3 Identifying Chromosomes

Chromosome	Characteristics
1, 2, 3	Very long; centromere in the center
4, 5	Very long; centromere not in the center
6, 7, 8, 9, 10, 11, 12	Long; centromere not in the center
13, 14, 15	Medium in length; centromere above the center
16, 17, 18	Medium in length; centromere at or close to the center
19, 20,	Short; centromere at or close to the center
21, 22	Short; centromere not in the center
Sex chromosomes	X is long; centromere in the center Y is short; centromere above the center

When you think you have properly identified and arranged all the chromosomes, tape or glue them to a single photocopy of Figure 18 on pages 98 and 99. Compare your final karyotype to the one shown on page 100. Do the two karyotypes match? They should. Note that this karyotype has one X chromosome and one Y chromosome, so the individual is a male.

FIGURE 18 KARYOTYPE WORKSHEET

1 – – –   2 – – –   3 – – –

6 – – –   7 – – –   8 – – –

13 – – –   14 – – –   15 – – –

19 – – –   20 – – –   21 – – –

4 – – – 5 – – –

9 – – – 10 – – – 11 – – – 12 – – –

16 – – – 17 – – – 18 – – –

22 – – – X – – – Y – – –

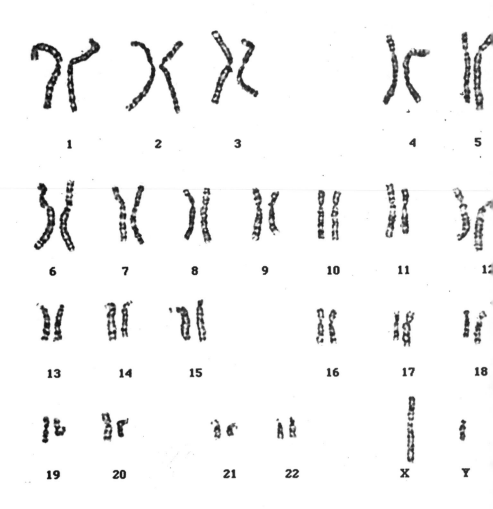

1   2   3   4   5

6   7   8   9   10   11   12

13   14   15   16   17   18

19   20   21   22   X   Y

Finished karyotype

Construct additional karyotypes using Copymasters 1, 2, and 3 (at the back of the book). Compare the karyotype you made from Copymaster 1 to your first karyotype. This individual is a female, so the karyotype has two X chromosomes and no Y chromosomes. How do karyotypes made from Copymasters 2 and 3 differ? Check the total number of chromosomes and look for missing portions of chromosomes. Correct karyotypes for the chromosomes spreads shown in Copymasters 1, 2, and 3 are provided in the Answers section at the back of the book.

THE SOCIAL IMPACT OF GENETIC TESTING

Tests that identify individuals at risk for genetic diseases have implications for the individual's family members and for society as a whole. Who should have access to the results? Do family members have a right to know? How about potential employers or health-insurance providers? There have already been cases in which people have lost jobs or promotions or been denied insurance coverage due to their genetic status. Finding ways to guarantee the confidentiality of gene test results is a major concern.

How do your relatives, friends, and classmates feel about this issue? Take a survey of thirty people to find out. How many of the people you interviewed would undergo genetic screening to determine if they had inherited a gene that predisposes them to a disease? Do they think other people should have access to the results? If so, who should and shouldn't be able to see the results? Should genetic testing be mandated on a wide-scale basis? Do research to learn more about the issues surrounding genetic testing.

Actual cases involving bioethics, particularly in genetic testing and therapy, often appear in newspapers, news magazines, and on television. Study these reports carefully and identify the issues involved in each case. Are there possible solutions to the problems? What would you do in the situations described?

UNDERSTANDING THE HUMAN GENOME

The Human Genome Project is a 15-year, multibillion-dollar quest to discover the location and function of every gene on our chromosomes. This is an extremely ambitious project. If the entire human genome were printed out, it would fill more than 1 million telephone-directory pages. Researchers in labs all over the world are working to create a very detailed "map" of our chromosomes. Other scientists are studying the function of genes. In most cases, a gene must be sequenced to determine its role in the body. Sequencing involves determining the exact order of nucleotide bases in the DNA fragment that contains the gene. When this project is completed, scientists hope to understand the hereditary instructions that make each of us unique.

A CLOSER LOOK AT DNA SEQUENCING

For nearly two decades, scientists have been using a method called chain termination sequencing, or the Sanger method, to determine the arrangement of nucleotide bases in a DNA molecule. In this procedure, four radio-labeled terminator substances are created. Each of the terminators is specially designed to respond to a different nucleotide base. For example, one terminator works only on adenine, another works only on thymine, and so on. Each terminator is added to one of four identical samples of DNA. As the DNA copies itself, the terminator stops the copying process each time it encounters the indicated base. The result is a series of fragments that all end with the same nucleotide base.

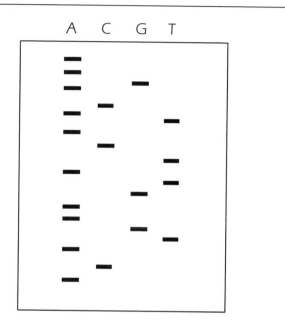

FIGURE 19 The order of bands on an X-ray film will provide the sequence of a strand of DNA that is complementary to the strand you are studying.

Next, each DNA sample is added to a separate well of an agarose gel slab and electrophoresed. The gel is then exposed to X-ray film. When the film is developed, it shows a banding pattern that represents the order of nucleotide bases in the original DNA sample. Based on what you know about DNA sequencing and electrophoresis, determine the nucleotide base sequence of the sample film shown in Figure 19.

CHAPTER 5

BIOTECHNOLOGY IN INDUSTRY

The biotechnology industry consists of companies engaged in the development of products and services in areas including agriculture, clinical diagnostics, food processing, energy production, and environmental conservation. These companies range in size from small entrepreneurial ventures to large, multinational conglomerates. In 1995, there were approximately 1,500 biotechnology companies in the United States. By the year 2000, the biotech industry is expected to generate more than 500,000 new job opportunities. This chapter explores some current biotech goods and services.

CRYOPRESERVATION

In 1776, Lazzaro Spallanzani, an Italian biologist, observed that when human sperm are cooled, they stop moving. Later, other scientists noted that when a cell is frozen, it dehydrates (dries out) because the water that makes up the cell's cytoplasm becomes inaccessible as it transforms to ice. The cooling process also increases the concentration of salts and other metabolites, which can damage the cell, as well as ice crystals, which can tear the cell's membrane.

In the early 1950s, biologists discovered that if they cooled human sperm in the presence of glycerol (glycerin) and stored the sample in dry ice, they could preserve the cells for a long period of time. When the sperm were rapidly thawed, the cells could fertilize an egg cell and a normal embryo would develop. Scientists named this process *cryopreservation*. Today, the technique is routinely applied to microbes, gametes, embryos, and cell cultures.

In the following project, you will cryopreserve the protist *Crithidia fasciculata*.

What You Need	
3 tubes with sterile Crithidia media	Soup can
	Freezer
Permanent marker	Plastic cup of water at 37°C
10 sterile transfer pipettes	
	Thermometers accurate between –30°C and 50°C
Actively growing Crithidia culture	
2 sterile 16 × 100-mm culture tubes	Microscope slides
	Coverslips
Glycerol (glycerin)	Compound microscope

Label the tubes with Crithidia media 1, 2, and 3. Using the sterile pipettes and the sterile technique described in the Technical Notes at the back of this book, transfer 5.0 mL of Crithidia media to Tube 1, 4.0 mL of Crithidia media to Tube 2, and 2.5 mL of Crithidia media to Tube 3. Add 2 drops of actively growing Crithidia culture to each of the two sterile 16 × 100-mm culture tubes. Using the same technique and fresh pipettes, add two drops of Crithidia culture to Tubes 1 and 3. Incubate all three tubes at room temperature for 24 hours.

Using a fresh pipette, add 1 mL of glycerol to Tube 2. Mix thoroughly by inverting the tube several times. Allow the glycerol to mix with the Crithidia medium for 60 to 90 minutes.

Add 2.5 mL of the liquid (cryoprotectant) in Tube 2 to Tube 3. Tubes 1 and 3 should continue incubating at room temperature. During this period, the cryoprotectant will begin to penetrate the cells.

Place Tubes 1 and 3 in the soup can (to ensure that they stay upright) and place the can inside a household freezer (the temperature should be about –20°C) for at least 48 hours.

Remove Tubes 1 and 3 from the freezer and place them in the plastic cup containing 37°C water. Using a standard laboratory thermometer, constantly monitor the temperature of the water in the plastic cup. Add hot water, dropwise by pipette, as necessary to maintain this 37°C temperature until the samples have thawed completely.

Add a few drops of the material in Tubes 1 and 3 to separate microscope slides. Place a coverslip over each sample and view these *wet mounts* under a compound microscope. View at 100X magnification. Was the cryoprotectant effective?

Doing More

- Prepare additional samples (with and without the cryoprotectant) to investigate viability—the number of motile cells—versus storage time. How does long-term storage (weeks or months) affect cell viability?

- Find out how other protists react to low-temperature storage. You might want to investigate cryopreserving Paramecium, Euglena, or other large ciliates. Use a 20% glycerin solution as the cryoprotectant and mix in a 50% cryoprotectant: 50% pondwater ratio.

- Try cryopreserving developing toad or frog eggs. Use sewing needles to separate egg masses. Use a 20% glycerin solution as the cryoprotectant; mix in a 50% cryoprotectant: 50% pond water ratio. Store the cells at 20°C (or at 79°C using dry ice) for 48 hours. Do the embryos develop normally after they are rapidly thawed?

RATE OF COOLING AND HEAT OF CRYSTALLIZATION

What You Need	
3 tubes with Crithidia media	Two $\frac{1}{2}$ × 2-inch cardboard strips
Permanent marker	2 thermometers accurate between –30°C and 50°C
2 sterile transfer pipettes	
Glycerol (glycerin)	Stapler
Test-tube rack	Freezer

Label the tubes with Crithidia media 1, 2, and 3. Using the sterile pipettes and the sterile technique described in the Technical Notes at the back of this book, transfer 5.0 mL of Crithidia media to Tube 1, 4.0 mL to Tube 2, and 2.5 mL to Tube 3.

Using a fresh pipette, add 1 mL of glycerol to Tube 2. Mix thoroughly by inverting the tube several times. Allow the glycerol to mix with the Crithidia medium for 60 to 90 minutes.

Add 2.5 mL of the liquid (cryoprotectant) in Tube 2 to Tube 3. Tubes 1 and 3 should continue incubating at room temperature. During this period, the cryoprotectant will begin to penetrate the cells. Place Tubes 1 and 3 in a test-tube rack.

Wrap a cardboard strip around each thermometer and staple the ends. Use these strips to position the thermometers inside Tubes 1 and 3 so that the thermometers do not touch the glass of the culture tubes. See Figure 20. Measure the temperature of each sample.

Place the test-tube rack in the freezer compartment of a refrigerator for 90 minutes. Continue to record the temperature of each sample at 5-minute intervals. Be sure to keep the freezer door open only long enough to take the readings.

Record all your data in a table in your lab notebook. The table should have the following headings: Time (in minutes), Temperature of Tube 1 (in °C), Cooling Rate of Tube 1 (Change in Temperature ÷ Amount of Time Lapsed), Temperature of Tube 3 (in °C), and Cooling Rate of Tube 3 (Change in Temperature ÷ Amount of Time Lapsed). In this case, the amount of time lapsed will always equal 5 minutes.

Graph your results. The y-axis (the vertical axis) should be labeled "Temperature (in °C)," and the x-axis

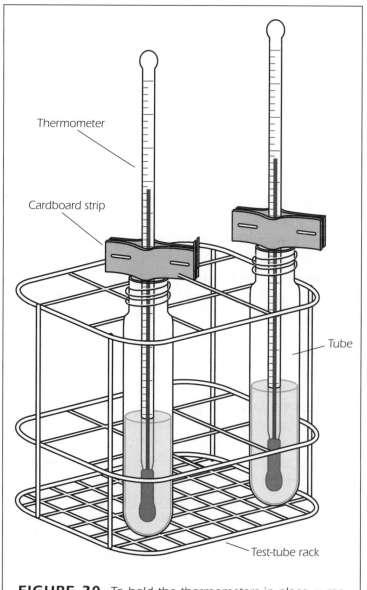

Thermometer

Cardboard strip

Tube

Test-tube rack

FIGURE 20 To hold the thermometers in place, wrap each in a $4 \times \frac{1}{2}$-inch strip of cardboard and staple each side.

(the horizontal axis) should be labeled "Time (in minutes)." Use the graph to determine the heat of crystallization—the amount of heat released when a liquid is transformed to the solid phase—for your samples. The majority of water present in cells usually transforms to ice crystals between $-2\,°C$ and $-5\,°C$.

DESIGNER DETERGENTS

An *enzyme* is a protein that speeds up a biochemical reaction. The enzymes produced by our bodies catalyze all kinds of reactions, from DNA replication to efficient removal of toxic carbon dioxide from blood cells.

Enzymes are also important ingredients of products we use every day. Enzymes obtained from microbes are added to laundry detergents, to remove stubborn stains such as grease, food, blood, and perspiration. At one time, the microbes were modified to produce enzymes that perform specific tasks. The modification process required exposing the microbes to strong chemicals or radiation. Today, bioengineering is used instead. The specific DNA sequence that codes for the desired enzyme is simply removed from the microbe and inserted into a second microbe that produces large quantities of the enzyme in a very short time.

Visit a local grocery store and look at all the detergents being sold. Find which enzymes are used in various detergent and stain-removing product brands (both liquids and powders). Although the contents labels on these products do not list specific enzyme names, you can use Table 4 to determine which enzymes are included in each product.

TABLE 4	Guide to Detergent Enzymes
Enzymes	**Use**
Proteases	Most widely employed. Remove protein stains such as grass, blood, egg, and perspiration.
Lipases	Remove grease and oil such as lipstick, frying fats, and salad oils.
Amylases	Remove starchy residues from spilled foods such as mashed potatoes, spaghetti, oatmeal, gravies, and chocolate.
Cellulases	Stain removers that "brighten" and "soften" fabrics.

WHICH DETERGENT IS BEST?

What You Need	
Nine 1-inch squares of cotton fabric	3 brands of detergent with different enzymes
Grease or oil	9 bowls
Starchy food	Warm water
Egg	

Stain three cotton fabric squares with grease or oil, three with a starchy food, and three with egg. Rub stains into the cloth squares.

Using directions on the label of each detergent, mix the proper ratio of warm water with detergent in the bowls. There should be three bowls with each brand of detergent. Immerse one grease-stained cloth square into each detergent type. Do the same for the cloth squares stained with starchy food and egg.

After 20 minutes, remove the pieces of cloth. Which detergent(s) removed each type of stain? What type of enzymes do you think are in each brand of detergent? Contact the manufacturer of each brand and ask if you are correct.

Doing More

When a colored garment made from cotton or cotton-blend fabrics is washed repeatedly, it tends to look "fluffy" and its colors become dull. These effects are due to the formation of microfibrils that become detached from the main fibers. The increased surface area reflects more light, making the fabric color appear duller.

Use a magnifying glass to compare the amount of microfibril formation on a new garment with one that has been washed many times in a nonenzyme detergent-based product. Wash each garment in a product that contains "brighteners" and "softeners" several times. Do you notice a difference? This product contains cellulases, which degrade the microfibrils, so the older garment should begin to look almost as good as new.

CHAPTER 6

BIOTECHNOLOGY
AND THE
ENVIRONMENT

An environment consists of the conditions under which an organism, or a group of organisms, lives. It includes all the factors that influence the physical location—soil type, climate, and other inhabitants. This chapter explores the ways scientists use living organisms to help maintain or restore the delicate balance required for organisms to survive.

NATURE'S RADIATION DETECTOR

In Japan, a *hybrid* of the common spiderwort plant, Tradescantia, is planted around many nuclear power stations in order to detect radiation leaks. When these plants are exposed to relatively high concentrations of ionizing radiation, the appearance of their stamen hair cells changes.

The common spiderwort is a tropical plant
with blue, purple, or white flowers.

Obtain a commercially grown spiderwort plant in bloom from a local florist or commercial grower and study its stamen hairs carefully with a magnifying lens. Pluck a stamen hair from a flower and observe a single cell through the magnifying lens. (Each individual plant cell looks like a bead.) You will be able to see the cell's nucleus if you make a wet mount of stamen hair cells and look at it under a compound microscope. These cells are unusually large.

To conduct this experiment, you will need the specially developed Tradescantia clone. After you receive it, follow the directions for forcing it to bud. Make arrangements through your physician, dentist, or a radiologist at a nearby hospital or clinic to x-ray one of the Tradescantia plants. Following X-ray exposure, place both plants in a sunny window.

Two days later, remove open flowers from the plants in the early morning. Using forceps, remove the stamens from each plant and place them on a wet paper towel. If necessary, use a dissecting needle to gently untangle the stamen hairs. View the hairs with a magnifying lens and count the number of chains of red cells (mutant sectors). Do this for both the treated and the control plant.

Over the next 2 weeks, count stamens from as many flowers as possible. Be sure to count stamens from an equal number of flowers from each plant. Each day, calculate the number of mutation events for both plants using the following formula:

$$\frac{\text{Number of Mutant Sectors Counted}}{\text{Number of Flowers Observed}} = \text{Average Number of Mutations per Flower}$$

Graph the average number of mutations for exposed and control plants over time. Are mutation events altered by the passage of time?

Doing More
Repeat the experiment using wild (not cultivated) Tradescantia (spiderwort) plants. Are the mutation frequencies of wild and commercially grown plants similar? Why? Consider the fact that the cloned Tradescantia are asexually propagated.

MAKING ICE THE BIOTECH WAY
The molecules that make up water are continually in motion, and the energy of this motion determines the water's temperature. If water is cold enough, its molecules slow down and align themselves in the latticed hexagonal array known as ice.

An *ice nucleator* is a substance that attracts water molecules and slows them down to form ice. Chemical compounds that can function as ice nucleators include sodium iodide, ammonium nitrate, and urea. In 1975, Steven Lindow, a University of Wisconsin plant pathologist, found that a protein growing on the cell wall of a naturally occurring bacterium, *Pseudomonas syringae*, can stimulate ice formation. When this ice-nucleating protein (INP) is mixed with water, the molecules of water attach to the protein in an arrangement that mimics the structure of an ice crystal. Marketed under the name Snowmax™, this protein is used at virtually all major ski resorts throughout the world.

What You Need	
Distilled water	Permanent marker
2 plastic containers	2 pipettes
INP granules	Two 4 × 4-inch squares of aluminum foil
Coffee stirrer	
	Access to a freezer

Add approximately 10 mL of distilled water to each plastic container. Add 3 or 4 INP granules to one container. Mix the solution with a coffee stirrer. Label the container "INP-treated Water." Label the other container "Plain Water."

Using a pipette, dispense rows of tiny drops of plain distilled water onto an aluminum foil square. Try to make the drops as equal in size as possible. Use the other pipette to dispense rows of tiny drops of INP-treated water onto the other aluminum foil square. Label both squares with a permanent marker.

Carefully place both these squares on a level surface in a household freezer. Observe the water drops on both squares for changes. Check at 5-minute intervals for a 30-

minute period, or until freezing occurs. Record your observations in your lab notebook. Did the INP-treated drops freeze faster?

STAYING ICE-FREE

THE ADVANTAGE OF PURITY

Pyrogen-free water contains very few impurities. In some cases, it will remain in its liquid state at temperatures as low as –40°C (–40°F). However, if a nucleator is added to this supercooled water, ice formation will occur spontaneously at any temperature below 0°C (32°F).

Obtain a small quantity of pyrogen-free water

What You Need	
Pyrogen-free water	2 medicine droppers
Two 4 × 6-inch pieces of aluminum foil	Tap water
	Access to a freezer

through your physician or local pharmacist. Use a medicine dropper to place several drops of pyrogen-free water on a piece of aluminum foil. Use a separate medicine dropper to place the same amount of tap water on the other piece of aluminum foil. Carefully place the foil pieces on a flat surface inside your kitchen freezer. Which type of water freezes first? Can you explain why?

SURVIVING THE WINTER

Insects, such as the gall moth *Epiblema scudderiana*, spend the winter inside spindle-shaped galls located along the stems of goldenrod plants. To do this, they empty their gut (to eliminate potential ice nucleators) and accumulate high levels of an internally produced antifreeze in their body tissues.

Try to locate moth galls on goldenrod stems in the fall or winter. Notice the position and orientation of each gall on the stem. Are the galls always high enough to be continuously exposed to winter conditions and yet not covered by snow? Record all your observations in your lab notebook. If possible, take some photos of the galls—especially after a snowfall.

Collect galls for detailed examination at home. Using a sharp knife or single-edged razor blade, slice the stems lengthwise through each gall. Carefully pry apart the stem and observe the silk cocoon constructed by the moth larva.

Use a pair of fine tweezers to remove a portion of this silk lining and place it on a clean, dry microscope slide. Place a drop of water on the silk and observe what happens. Do research to learn about the proteins fibroin and sericin. What are their characteristics? Can you deduce the last part of the over-wintering freeze-avoidance strategy?

TOO MUCH SUN

Ultraviolet radiation—specifically the UV-B component (280 to 320 nm)—can damage the DNA of most organisms. Although many creatures have evolved the ability to repair this damage, in some cases, the damage is too great and the affected cells die.

One strain of baker's yeast (Strain G948-1C, *Saccharomyces cerevisiae*) has been specifically altered in order to inhibit its DNA repair mechanisms. As a result, this organism is a perfect candidate for radiation sensitivity experiments.

What You Need

Transfer chamber	Box of toothpicks (with a small hole in one corner)
Paper towels	
Diluted household bleach (1 part bleach to 9 parts water)	6 sterile cotton-swab applicators
Saccharomyces cerevisiae, strain G948-1C	25 mL baby-food jar containing sterile water
5 prepared petri dishes of YED agar	Masking tape
	Four small squares of paper

Caution: Wash your hands using antibacterial soap before and after performing this experiment.

Sterilize your transfer chamber by wiping down the inside surface of the box with a paper towel and diluted bleach. Place a few clean paper towels inside the chamber and add the following materials : yeast, a petri dish, and the box of toothpicks.

Working inside the transfer chamber, remove a toothpick from the box with one hand and raise the lid of the petri dish with the other hand. Use the broad end of the toothpick to pick up yeast cells and transfer them to the surface of a petri dish. Carefully apply the tip of the toothpick to the surface of the agar and streak it. Close the lid of the petri dish.

Incubate the petri dish for 2 or 3 days at room temperature. Remember to incubate the petri dish upside down to prevent condensation from dripping onto growing colonies. Look for emerging white colonies growing up from where you streaked the dish's surface. (You can store the "original" yeast strain culture in the refrigerator—if your parents approve—for up to 9 months.)

Resterilize your transfer chamber by wiping down the inside surface of the box with a paper towel and diluted bleach. Place a few clean paper towels inside the chamber and add the following materials: the petri dish with yeast, the four unused petri dishes with YED agar, sterile swab applicators, and the baby-food jar containing sterile water.

Raise the cover of the petri dish containing the yeast and rub the cotton swab over the yeast. Open the jar containing sterile water. Swirl the swab in the water to wash off yeast cells. Continue to add yeast to the water until it turns cloudy. Be sure to use new swab applicators each time you add more yeast.

Dip a clean sterile swab into the yeast suspension in the baby-food jar. Withdraw the swab and carefully streak the surface of a new petri dish. Rub the swab across the petri dish in one direction, then turn the dish 90 degrees and rub the swab across again. This technique should guarantee that the entire surface is covered with yeast. Don't worry about excess moisture—it will soak into the agar. Tape the cover of the petri dish closed. Repeat this procedure using the three remaining petri dishes.

Expose all four dishes to sunlight for 10 to 30 minutes. The dishes should be perpendicular to the sun's rays to achieve maximum exposure. To make sure that they are, affix a small paper square to the top of each dish and observe the shadow cast on the agar's surface. The smaller the shadow, the more perpendicular the position of the dish. Ideally, there should be no shadow at all.

Incubate test petri dishes for 2 to 4 days. Be on the lookout for signs of colony growth. Disinfect dishes before disposing of them.

Doing More

Try the following exposure conditions and record your results (colony growth) in your lab notebook:

- Compare the effectiveness of various commercial sunblocks by applying these products to the upper surface of some petri dishes.

- How does the angle of the sun affect the amount of UV-B falling on the cells? Do exposures at noon result in greater cell death than exposures when the sun's angle is lower on the horizon?

- On which day of the year (assuming a clear day) would you expect the lowest radiation exposure at noon in the Northern Hemisphere?

- What happens if you expose a petri dish to the sun for 2 to 4 minutes a day (at the same time each day) for several days?

- If possible, compare exposed dishes at various elevations such as sea level versus a mountaintop. Does elevation affect your results?

OIL AND WATER DON'T MIX

Crude oil is a complex substance made of many different components. When crude oil is refined, these components are separated out and used in a variety of ways. The world relies on crude oil for heating oil, automobile gasoline, plastics, and many other materials.

The refining process involves fractional distillation. In this process, the components are collected as they evaporate out of the original mixture. At the end of the process, lubricating oil and asphalt are left behind. Table 5 lists all

TABLE 5 Components of Crude Oil	
Fraction	Distillation temperature (°C)
Gas	< 20
Petroleum ether	20–60
Naphtha	60–100
Gasoline	40–205
Kerosene	175–325
Lubricating oil	Not available (liquid)
Asphalt	Not available (solid)

the constituents of crude oil and the temperatures at which they vaporize.

Since most of the world's oil fields are in the Middle East, tankers must be used to transport oil to other parts of the world. Once in a while, an oil tanker is in an accident and some of its oil cargo leaks into the ocean. An oil spill can be an ecological disaster.

A single gallon of spilled oil can spread to cover 4 acres of water! Within days, the volatile components of the oil evaporate, leaving behind a highly viscous, *immiscible* substance that clings to marine organisms. After 3 months, 15 percent of the original volume of oil remains in the form of tar balls. The rest of the oil has been degraded by a variety of micro organisms or radiant energy from the sun. This tar washes up on distant shores months after a major oil spill. In the projects that follow, you will use an oil-degrading microbe to treat a miniature oil spill.

BIODEGRADATION
OF OIL

<table>
<tr><th colspan="2">What You Need</th></tr>
<tr><td>Distilled water</td><td>Compound microscope</td></tr>
<tr><td>Culture tube with cap</td><td>Bottle of bacteria stain</td></tr>
<tr><td>2 medicine droppers</td><td>(1% crystal violet, 1%
methylene blue, or</td></tr>
<tr><td>Refined motor oil (not
used motor oil)</td><td>0.25% safranin stain)

Paper towels</td></tr>
<tr><td>Oil-degrading microbe
broth culture</td><td>Access to a sink</td></tr>
<tr><td>Microscope slides</td><td></td></tr>
</table>

Caution: Wear protective polyethylene gloves and safety goggles to minimize the risk of any contamination.

Add 5 mL of distilled water to a culture tube. Using a medicine dropper, add 2 to 3 drops of refined motor oil to the tube of water. The oil should form a thin layer on top of the water.

Using a clean medicine dropper, add 5 to 10 drops of an oil-degrading microbe broth culture to the tube containing the motor oil. Screw the cap on the culture tube and invert it several times to ensure good mixing and to simulate wave intermixing. Loosen the cap of the tube and incubate it vertically at a temperature as close to 30°C as possible. Wash your hands thoroughly. Illustrate and describe the oil layer in your lab notebook.

Continue to observe the tube every day for several weeks. Invert the tube 2 to 3 times each day to simulate wave action and promote intermixing. Carefully note changes in your lab notebook. How long does it take to degrade the oil sample?

After a week or so, use a medicine dropper to remove a tiny portion of the oil layer. Examine the oil under a compound microscope. Compare this material with a sample of fresh, refined motor oil. Does the cloudiness of the sample increase over time?

Make a stained microscope slide preparation to find out what's happening at the microscopic level. (Follow the instructions listed in the Crown Gall Tumors: A Closer Look project in Chapter 3.)

BIOREMEDIATING AN OCEAN OIL SPILL

What You Need	
Distilled water	3 medicine droppers
Baby-food jar with cap	Oil-degrading microbe broth culture
Refined motor oil (*not* used motor oil)	Powdered plant fertilizer

Caution: Wear protective polyethylene gloves and safety goggles to minimize the risk of contamination.

Add distilled water to a baby-food jar until it is half full. Using a medicine dropper, add enough refined motor oil to form a spill on the water's surface. Using another medicine dropper, add 25 to 30 drops of oil-degrading microbe broth culture to the spill. Cover as much of the oil slick's surface as possible.

Using your thumb and forefinger, sprinkle fertilizer over the spill. (Such nutrient material is routinely applied to spills to enhance microbial growth, absorb oil, and act as a growing surface for the microbes.)

Incubate the jar with its cap loosely attached at 30°C. Agitate the mixture by using a medicine dropper to blow bubbles into the water at least twice daily. (This simulates wave action.) Wash your hands thoroughly.

Observe the tube every day for several weeks. Carefully note changes in your lab notebook.

Doing More

- Compare these bioremediation measures with the application of liquid detergents to similar spill scenarios. Which method tends to be more effective over time?

- Design a project that investigates the effects of temperature on oil degradation.

- Design a project that simulates a shoreline spill by including sand and pebbles in a petri dish along with oil and water.

COMPOSTING: RECYCLING NATURE'S WAY

Composting is the biological decomposition of organic waste materials under controlled conditions. It was first used by the ancient Romans more than 2,000 years ago to improve the quality of their soil. Today, composting is also used as a way of managing wastes. Composting food and yard waste reduces the total volume of material that ends up in landfills and incinerators. In fact, composting has been classified as a form of recycling. In the past decade, many communities have turned to composting as an environmentally sound and economically feasible solution to waste-disposal problems.

Composting takes place continuously in nature, as plants and animals die and decompose to form humus. During the process of decomposition, bacteria, fungi, and a few other organisms break down plant and animal remains into simpler components. As a result, the nutrients contained in the bodies of the plants and animals are released into the soil, where they can be used by living plants.

What You Need	
Organic wastes from home and yard	Garden hose Spade

Clear a moderately sunny corner of your yard and begin to make two parallel piles of organic waste material, such as vegetable and fruit peelings, eggshells, coffee grounds, tea leaves, grass clippings, apple cores, leaves, wood ashes, hay, and brush. As the piles grow, try to maintain a 2:1 mixture of carbon to nitrogen. Generally, fresh green plant materials are high in nitrogen, while dried leaves, hay, and brush are high in carbon.

During the first 6 weeks, gently moisten the waste material with a garden hose every day, then shred and turn the material over with a spade. (Do not use the hose on rainy days. On very hot days, turn the pile twice to eliminate an odor buildup.)

Let the material sit undisturbed for the next 6 months. You will know that the compost is ready to use when it is dark brown, crumbly, and smells like soil. In most cases, a pile started in the summer or fall should be ready for use the following spring.

Doing More

A compost pile is really a miniature recycling factory. Microbes in the pile repackage organic and inorganic nutrients locked up within the cellular structure of household and yard waste into a form that plants can use.

- Monitor the temperature at various depths of the compost pile. Are all levels of the pile equally warmer or colder than the air? Do temperature changes increase a few hours after the pile is turned?

- After the initial 6-week period, bury some 1 × 3-inch glass microscope slides at various locations within the pile. Mark the location of each slide with a stick labeled with the date. Do not disturb the pile for 6 to 8 weeks. Then, carefully remove the slides and stain them according to the procedure described in the Crown Gall Tumors: A Closer Look project in Chapter 3. Observe the slides under high-dry magnification (430X) of a compound microscope. In your lab notebook, list the types of organisms that live in different parts of the pile. Why do you think different types of microbes live in different parts of the pile?

- Use a soil-test kit (available at lawn-care centers) to analyze the nutrient content of your compost. Does it have measurable amounts of phosphorus, nitrogen, and other nutrients? What is its pH?

CHAPTER 7

BIOTECHNOLOGY IN THE COURTROOM

O n the morning of November 22, 1983, the body of a 15-year-old girl was discovered just outside the grounds of a mental hospital in Narborough, England. She had been brutally raped and strangled. Although it was not immediately evident who committed the crime, a very important piece of evidence was left behind. Police extracted a sample of the rapist's DNA from semen found at the crime scene. Three years later, the same DNA results were obtained from a semen sample recovered from another 15-year-old victim. The DNA samples eventually turned out to be the clues that solved both crimes.

Shortly after the second murder, the local police arrested a man who worked at the mental institution where the first body was found. He confessed to the second murder, but denied any knowledge of the first one. Because the police were fairly certain that both murders had been committed by the same person, they needed to find some way of proving that the suspect had killed the first victim.

About the same time, Alec Jeffreys, a geneticist at the University of Leicester in England, was studying inherited variation within genes and among individuals as well as the evolution of present-day genes. His work involved looking at a genetic peculiarity known as the *intron*. Introns are sequences of "nonsense" or "junk" DNA. In other words, they do not code for a specific protein.

Jeffreys noticed that some introns are made up of the same repeating DNA base pair sequences and that the number of repetitions varies from person to person. For example, in one person a particular base pair sequence may repeat ten times, while in another person the same sequence may repeat twenty-five times. These repetitive intron fragments are called restriction fragment length polymorphisms (RFLPs). Although the DNA sequences of genes—fragments that code for a protein—are fairly constant from person to person, introns are not. RFLPs are unique to each person, except in the case of identical twins.

Jeffreys realized that these variable DNA base pair sequences could be separated using gel electrophoresis and matched with complementary radioactive probes to create an individual-specific DNA banding pattern that would distinguish one person from the next. Because every person has a distinct RFLP banding pattern, Jeffreys called this technique DNA fingerprinting.

Narborough police asked Jeffreys to verify the suspect's confession by comparing his DNA fingerprint to the one obtained from the semen. The tests proved that the suspect had not committed the murders. They also confirmed that both crimes had been committed by the same man. In a desperate move, police asked all male residents to provide a sample of their blood for DNA analysis. After several months of testing, a match was found and a man by the name of Colin Pitchfork was arrested.

Today, DNA fingerprinting is routinely used to test samples of blood, hair, saliva, and semen left at crime scenes. The results are compared to those of a suspect to establish the individual's guilt. DNA fingerprint results are not considered irrefutable evidence in court. There is a slight chance that two individuals could have the same RFLP banding pattern. When the medical personnel who do the testing appear at a trial, they tell the jury the statistical probability that two matching samples (the one from the crime scene and the one from the suspect) came from the same person. This figure is based on the number of known RFLPs that exist in a given population. In some cases, a match can be accurate to within 1 in 10 billion people. Since that's twice the current human population, it is extremely likely that the suspect did commit the crime.

Although DNA fingerprinting is a very accurate technique, some lawyers argue that it can never be accurate enough to prove guilt beyond a reasonable doubt. Many legal challenges have been raised—most recently in the highly publicized case of O. J. Simpson—concerning DNA evidence. Because the techniques used to produce DNA fingerprints are constantly being improved, the likelihood of misidentification is much lower today than it was a few years ago.

DNA fingerprinting has many other potential applications. Since half of a person's genome comes from each parent, DNA fingerprinting can be used to determine familial relationships. It can also be used to track hereditary diseases and identify the best matches for organ transplants. This procedure can even be used to ascertain the level of inbreeding in endangered animals like the cheetah and to determine how newly discovered species are related to other organisms on Earth.

DNA WHODUNIT
WHO STOLE THE GOODS?

Crime investigators were called to the scene of a burglary. Mr. Jones had come home to find someone burglarizing his apartment. After a brief struggle, the burglar ran across the room and escaped through an open window. In his hurry, the burglar accidentally tore his shirt and cut his arm on the window frame. The crime investigators were able to remove small pieces of blood-stained clothing from the window frame. The material was immediately sent to the forensics lab to be analyzed. The results showed that the sample had blood type A, which also happened to be Mr. Jones's blood type.

After the crime investigators carefully reviewed all of the available evidence, they apprehended three suspects. To everyone's surprise, all three suspects had blood type A and matched the general description given by the victim. The only technique that could possibly provide more conclusive evidence about who actually committed the crime was DNA fingerprinting.

You have been chosen to perform the tests and interpret the results. Remember, if you make a mistake, an innocent person might go to prison. Study the illustrated autoradiogram shown in Figure 21. Compare the DNA banding pattern of the three suspects with patterns from samples at the crime scene, as well the victim's DNA. Do any of the patterns match? Were any of the suspects involved in the crime?

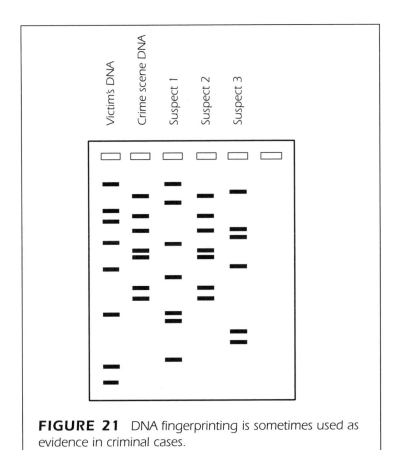

FIGURE 21 DNA fingerprinting is sometimes used as evidence in criminal cases.

WHO'S THE FATHER?

Before DNA fingerprinting was developed, paternity testing was based on blood type. This is because an individual's blood type is based on the blood types of his or her parents. Since many people have the same blood type, this type of test can only be used to exclude possible biological parents, but not to actually prove who the parents are. DNA finger-printing, on the other hand, can prove familial relationships with a high degree of certainty.

Figure 22 is an illustration of an autoradiogram show-ing RFLP banding patterns of a child, the mother, the alleged father, and a combination of the child and the alleged father. Based on your knowledge of DNA finger-printing, determine if the alleged father is really the bio-logical father of the child. (Remember that half of a per-son's genome comes from each parent.)

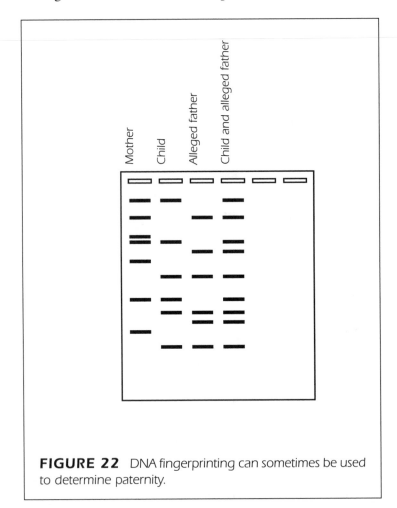

FIGURE 22 DNA fingerprinting can sometimes be used to determine paternity.

COPYMASTERS

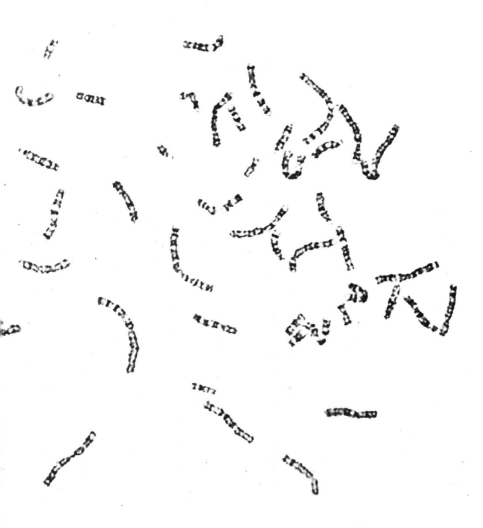

Copymaster 1
Chromosome Spread

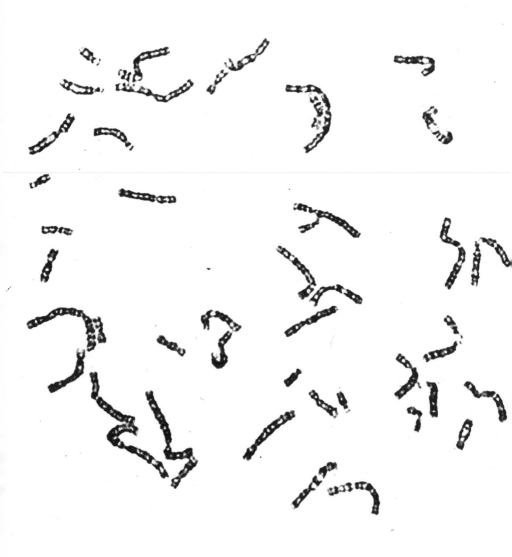

Copymaster 2
Chromosome Spread

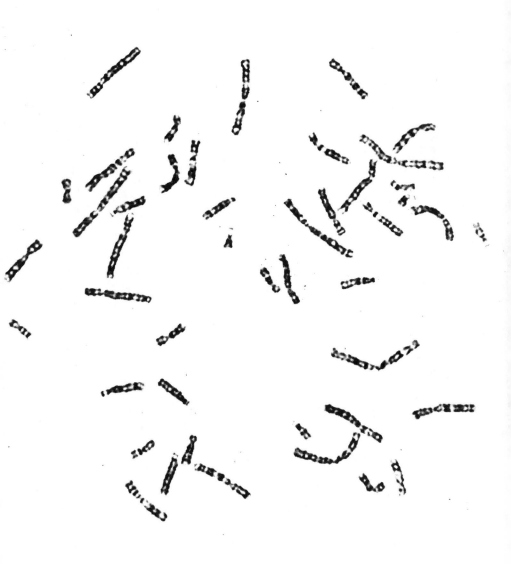

Copymaster 3
Chromosome Spread

ANSWERS

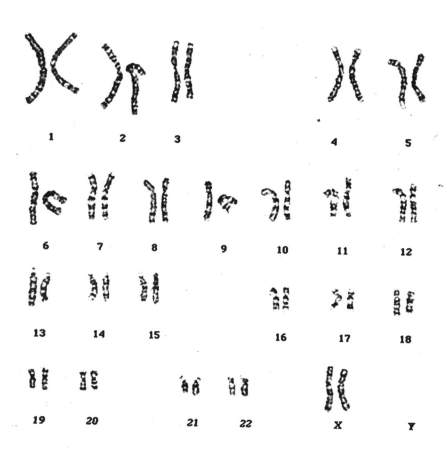

Karyotype 1   Normal Female (46, XX)

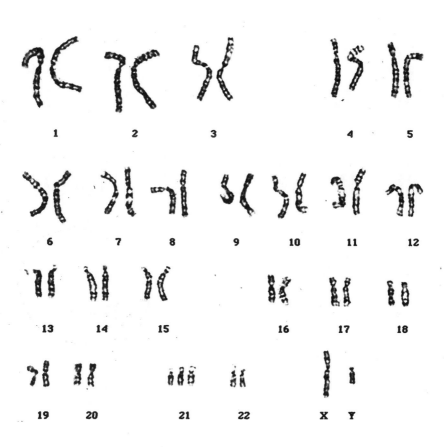

Karyotype 2   Down Syndrome (47, XY, +21)

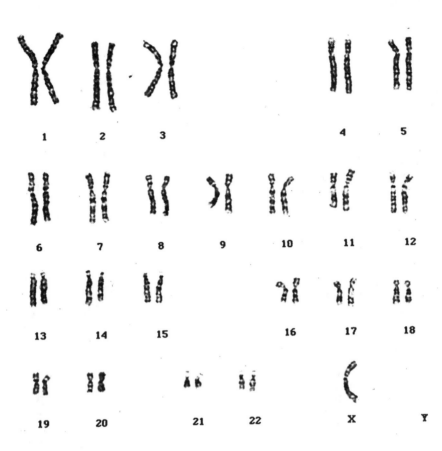

1   2   3   4   5

6   7   8   9   10   11   12

13   14   15   16   17   18

19   20   21   22   X   Y

Karyotype 3   Turner's Syndrome (45, X)

Missing or Extra Chromosomes

Down syndrome is a chromosomal disorder that occurs when an individual inherits an extra copy of chromosome 21. The incidence of this disorder is approximately 1 in 600 live births. It causes mental retardation, distinct facial features, and heart defects.

Another common chromosomal disorder is Turner's syndrome. Individuals with this disorder have only one X chromosome and no Y chromosome. The incidence of this disorder is approximately 1 in 5,000 live female births. Because these individuals have no Y chromosome, they are females. However, they lack functional reproductive organs and have a range of abnormalities, including short stature, renal and cardiovascular anomalies, and lower-than-average IQ.

TECHNICAL
NOTES

TIPS FOR PREPARING SOLUTIONS

Equivalents
- One level teaspoon = 5 g or 5mL
- One cup = 240 mL
- One fluid ounce = 30 mL
- One drop = 0.1 mL

Molar Solutions
A molar solution contains the molecular weight in grams of a substance made up to 1 liter with distilled water. Molecular weight expressed in grams is called the gram-molecular weight or mole. For example, a molar solution of oxalic acid $(COOH)_2 \cdot 2H_2O$ would consist of 126 (gram-molecular weight) grams of the substance in 1 liter of distilled water. Normal solutions (N) are generally considered essentially equivalent to molar solutions for the reagents used.

Percentage Solutions
When preparing percentage solutions, it should be clear whether the percentage is determined by weight or volume expressed as follows:

w/v = weight in grams in a 100-mL volume solution

v/v = volume in milliliters in a 100-mL volume solution

MAKING AGAROSE GEL
Agarose is a natural polysaccharide derived from seaweed. To make a gel, you can either melt a 2% precooked formula purchased from a supply house or prepare it yourself.

Do this procedure under adult supervision. Mix 2 g of agarose with 100 mL of dilute running buffer (1X concentration) in a container that can be heated on a kitchen stove. Bring the solution to a boil. Stir until the agarose is fully dissolved in the buffer. Allow the agarose to cool down a bit before pouring it into a gel casting tray. If you do not need the gel right away, use a funnel to pour it into a clean, capped plastic bottle. If refrigerated, the gel can be stored for many weeks.

RUNNING BUFFER
Commercially available TBE (Tris Borate EDTA) electrophoresis running buffer works best and is available from suppliers. A simple electrolytic solution of 5 g sodium chloride and 240 mL of distilled water is adequate.

FERMENTATION MEDIA
Sterile tryptic soy broth media can be purchased from a biological supply house. However, an alternative media can be prepared using soybean flour as follows:

In a clean pot, mix 4 teaspoons of soybean flour and 1 tablespoon of corn syrup in 500 mL of distilled water. Place the mixture over low heat and bring to a boil while stirring. Cover the media with aluminum foil and allow it to cool to about 40°C before transferring it to the fermentation vessel.

WORKING WITH BACTERIA
Bacteria are all around us—and in us! They are tiny biochemical factories that reform and assimilate organic matter. Scientists have estimated that the number of bacteria found in a pailful of garden soil is greater than the number of humans that have ever lived.

Although very few of the more than 10,000 bacteria that have been identified by biologists are harmful to humans, bacteria should always be handled with care. Never work with bacterial cultures if you are sick, are under a doctor's care, or have open sores or cuts. Also, avoid breathing the dust. These simple precautionary measures will minimize the risk of becoming infected by a harmful microorganism.

Before working with bacteria, wash your hands thoroughly, and wipe down work surfaces with disinfectant solution

(1 part laundry bleach to 9 parts tap water) or use a commercial disinfectant solution such as Lysol™. As you work, keep the area clear of unnecessary objects.

Because packaged toothpicks are sterile, they make great transfer tools. With practice, you will be able to inoculate a petri dish's media without breaking the surface.

Use sterile cotton-swab applicators rather than toothpicks when you want to cover the entire surface of the petri dish with bacteria.

When you have completed your work with a bacteria sample, all materials that have come in contact with the the bacteria must be decontaminated prior to disposal. If an autoclave or a pressure cooker is not available, soak toothpicks and swabs in laundry bleach before throwing them away. Add laundry bleach (or Lysol) to the media surface of used petri dishes until the media is covered. Once the disinfectant solution is absorbed by the agar, it is safe to discard the petri dishes.

DNA STAINING

Caution: Wear safety goggles, rubber gloves, and an apron or smock when working with stains.

Place the gel in a suitable plastic container. Pour DNA stain over the gel until it is completely submersed. Allow the gel slab to stain for about 1 hour. (The time required for staining will vary depending on the type of stain used.) Pour off stain and wash gel in successive changes of tap water over a 24-hour period. Place the gel in a sealable plastic bag. You can examine the gel on a light table or hold it up to a light source.

WORKING WITH FRUIT FLIES

You can obtain fruit fly culture and dry "instant" culture media from your science teacher or suppliers. *D. mojavensis* has larger salivary glands than *D. melanogaster* and is thus preferable for beginners. Third-instar larvae can be obtained directly from shipped cultures.

Drosophila cultures intended for salivary-gland preparations should be kept at room temperature—68 to 72°F (20 to

22°C). Follow package directions when working with the instant medium. You will also need a 4 × 1-inch clear plastic vial and polyurethane foam or nonabsorbent cotton plug. Third-instar larvae should be selected when they crawl out of the medium onto the sides of culture container. Choose organisms that look fully developed, but still have soft cuticles. Place culture vials in a refrigerator for 15 to 20 minutes to immobilize flies.

WORKING WITH PROTISTS

An infusion is a liquid broth that is rich in nutrients. It can be used to grow various types of microorganisms. You make an infusion just like you make sun tea—except without the sun or the tea! Place a handful of finely chopped dried-grass clippings in a mayonnaise jar. Add distilled or bottled spring water (available at grocery stores) to the jar and set it on a windowsill that receives moderate sunlight.

To culture the protists, obtain a "starter" inoculum of Chlorella or Euglena through your science teacher. Put 15 g of chopped Timothy hay into a quart-size canning jar. Place the jar in a saucepan and add boiling water to fill the jar to within 1 inch from the top. Allow it to cool undisturbed, then cap. Store the jar in a refrigerator. Add the "starter" inoculum to a jar of hay infusion medium. Set the jar in a sunny window, and wait for it to turn pea green.

You can also obtain a "starter" inoculum of Paramecium through your science teacher. Add 15 g of chopped Timothy hay, a beef bullion cube, and a pinch of yeast into a quart-size canning jar. Place the jar in a saucepan and add boiling water to fill the jar to within 1 inch from the top. Allow it to cool undisturbed, then cap. Fill a medicine dropper with Paramecium culture and add it to the jar containing the hay infusion. Also add 2 to 4 rice grains. Gently cover the jar and place it on a shelf that is not exposed to light. A flourishing culture should appear in about 1 week.

An actively growing sterile Crithidia culture can be purchased from a supply company. It can be maintained for up to

10 days at room temperature prior to culturing on Crithidia medium using the sterile technique described below.

STERILIZATION

Sterilize materials such as forceps, flasks, and rubber stoppers by placing them in a pressure cooker at a temperature of 121°C (250°F) (15 PSI pressure) for approximately 15 minutes. If a pressure cooker is not available, boil equipment in a pot for about 30 minutes. Use tongs to remove sterilized items. After sterilizing flasks and other containers, place aluminum foil over their openings to prevent contamination by airborne microbes.

It is also possible to chemically sterilize equipment. Let the items soak in a dilute bleach solution (1 part bleach to 3 parts water) for about 10 minutes. Small articles, such as forceps, may be surface-sterilized by the application of alcohol from an alcohol pad.

STERILE TECHNIQUE

The *sterile technique* involves the careful manipulation of sterile instruments and growth media during culture inoculation or transfer. The procedure prevents contamination by bacteria in the air and on surfaces. Practice the procedure before working with sterile materials. Keep the following guidelines in mind:

- All instruments, growth media, and working surfaces should be sterile.
- Try to work inside a transfer chamber to minimize contamination by airborne microorganisms.
- Keep the transfer area free of unnecessary objects and equipment.
- Wear clean clothing.
- Avoid hand-to-mouth operations.
- When transferring liquids, be sure to remove caps and place them upside down (threads facing up) during transfer operations, unwrap sterile pipettes carefully, and do not insert pipettes into culture tubes any farther than is absolutely necessary.

STERILE WATER DISPENSER

You can use 4-oz. (120-mL) baby bottles as sterile water dispensers. To sterilize a bottle, remove the cap and nipple and place all three pieces in a saucepan of boiling water for 15 minutes. Remove the pieces with kitchen tongs. Place the sterilized bottle in another saucepan and fill the bottle with distilled water. Push the nipple up through the cap ring and loosely screw this assembly onto the bottle. Add tap water to the saucepan and bring to a boil for 15 minutes. Allow the bottle to cool until it can be comfortably handled, then tighten the cap.

GLOSSARY

allelopathic—the influence or interference of a plant through the production of certain chemical agents on the growth of other nearby plants or microorganisms.

assay—a laboratory procedure for measuring biological or chemical activity or identifying a reaction.

bioengineering—the use of genetic engineering technology in the biosynthesis of economically important compounds.

bioreactor—a device in which living cells are mixed with nutrients and grown in a sterile and controlled environment to produce a desired product. It is also called a fermenter.

centromere—the point at which the two halves, or chromatids, of a chromosome are attached.

chromosome—a rod-shaped structure consisting of DNA and protein. Chromosomes are located in the nucleus of all eukaryotic cells.

chromosome spread—an image of all the chromosomes isolated from a cell before they are arranged into a karyotype.

clone—a genetically identical cell, group of cells, or individual derived from a single cell.

cryopreservation—a method of sub-zero freezing used to preserve the viability of cells and microorganisms.

control—a test carried out to provide the norm or standard against which experimental results can be compared.

culture—the deliberate, propagated growth of an organism—usually a microorganism—on or within a solid or liquid medium, usually as a result of prior transfer (inoculation) and incubation.

denature—to separate the two strands of a DNA molecule using heat, a change in pH, or irradiation.

diabetes—a disease that occurs when an individual's pancreas secretes abnormally low concentrations of insulin.

enzyme—a complex protein that catalyzes a biochemical reaction.

fermentation—a wide range of biological processes in which living organisms or enzymes are combined with nutrients in a controlled environment to produce commercially useful products.

gel electrophoresis—a process in which the molecules of a substance are separated according to their molecular weight.

gene—a section of a chromosome containing the information for a particular trait.

genetic engineering—a technology used to produce artificially genetically modified organisms, microorganisms, cells, and plants by the substitution or addition of a new gene from a different organism.

genetic marker—a piece of DNA or gene with known properties. It is used as a control to identify unknown genes.

herbicide—a chemical used to kill or control the growth of plants.

heterozygous—inheriting a different form of a gene (allele) for a given trait from each parent.

homologous pair—a pair of chromosomes that contain the same sequence of genes, but are derived from different parents.

homozygous—inheriting the same form of a gene (allele) for a given trait from both parents.

hybrid—heterozygous; an organism that has inherited a different form of gene (allele) for a given trait from each parent; any cross-bred animal or plant.

ice nucleator—a particle in the water that starts the freezing process by attracting water molecules and slowing them down.

incubate—to place a living organism or an organism culture in a specific environmentally controlled chamber, at a given environmental temperature, for a specific period of time. The goal of the incubation process is to grow a new population of a specific organism.

inoculation—using sterile technique to deposit material (the inoculum) on or into a sterile growth medium or a living organism. Generally, inoculation is performed to start an organism culture that is subsequently incubated to produce a large population of organisms.

instar—in some insects, the stage of development between molts.

immiscible—incapable of mixing.

intron—a "nonsense" DNA base sequence (also referred to as junk DNA) that does not carry any useful genetic information.

karyotype—a visual display of an individual's chromosomes. It is arranged as homologous pairs in order of chromosome size, shape, and centromere location.

laws of dominance—the tendency of certain stronger alleles (dominant) to conceal the trait of the weaker corresponding alleles (recessive).

legume—a type of fruit derived from a single carpel that splits down both sides at maturity. Examples include: peas, beans, and clovers.

medium (media)—a mixture of nutrients needed for cell growth.

mutation—an alteration in the chemical structure of DNA. The resulting change may affect the characteristics of an individual cell or an entire organism.

nucleotide base—a structural unit of a nucleic acid (DNA or RNA) consisting of three parts: a sugar, a phosphate group, and a base.

plasmid—a small ring of DNA in bacteria.

root nodule—a structure formed on the roots of leguminous plants containing nitrogen-fixing bacteria.

sequence (DNA)—the order of nucleotide bases in a DNA molecule.

sporulate—to go into a protective stage, due to a reduction of available nutrients. The term generally pertains to bacteria.

sterile technique—a laboratory method that involves the careful manipulation of sterile instruments and growth media during culture inoculation or transfer. The goal of the method is to prevent contamination.

totipotent—having the ability to develop into a complete organism. Some cells are totipotent.

wet mount—a temporary microscope slide preparation using water as the mounting media.

SCIENCE SUPPLY COMPANIES

Most supply companies will supply materials if a mail order is accompanied by a check, or credit card number when ordering by phone, FAX, or through a Web site. You can borrow catalogs from your teacher to get information about order numbers and prices. Use the list below with the Materials Guide section to find out which companies carry the items you need. In most cases, you should coordinate your order through a science teacher.

1 **Carolina Biological Supply Company**
York Road, Burlington, NC 27215
Phone: (800) 547-1733 FAX: (800) 222-7112
e-mail: cbass@carolina.com
Website: http://www.carosci.com

2 **Connecticut Valley Biological
Supply Company**
Valley Road, Southampton, MA 01073
Phone: (413) 527-4030 FAX: (800) 523-2109

3 **Science Kit and Boreal Laboratories**
777 East Park Drive, Tonawanda, NY 14150
Phone: (800) 828-7777 FAX: (800) 828-3299
Website: http://www.scikit.com

4 **WARD'S Natural Science Establishment**
5100 W. Henrietta Rd, P.O. Box 92912,
Rochester, NY 14692
Phone: (800) 962-2660 FAX: (800) 635-8439
e-mail: customer_service@wardsci.com
Website: http://www.wardsci.com

MATERIALS GUIDE

Most of the materials required for projects in this book are available at local stores. Microorganisms, laboratory apparatus, and reagents can be purchased from the science supply companies. The following table tells you which companies carry the materials you will need. Generally chemicals and microorganisms must be purchased by your science teacher.

CHAPTER 1

Materials	Suppliers
Agarose	1, 2, 3, 4
Electrophoresis chamber (battery-powered)	4
Fermentation Introductory Kit	4
Running buffer	1, 2, 3, 4

CHAPTER 2

Materials	Suppliers
Agarose	1, 2, 3, 4
DNA Extraction Kit	4
DNA stain	1, 2, 3, 4
Drosophila cultures and culture media	1, 2, 3, 4
Electrophoresis chamber (battery powered)	4
Glycerin	1, 2, 3, 4
Lactose	1, 2, 3, 4
Loading dye	1, 2, 3, 4

CHAPTER 3

Materials	Suppliers
Agrobacterium tumefaciens	1, 2, 3, 4
Bacillus strains	1, 4
Bacteria stains	1, 2, 3, 4
Callus tissue	1, 4

CHAPTER 4

Materials	Suppliers

CHAPTER 5

Materials	Suppliers

CHAPTER 6

Materials	Supplier

CHAPTER 7

Materials	Supplier

RESOURCES

BOOKS

Bishop, Jerry. *Genome: The Story of the Most Astonishing Scientific Adventure of Our Time—The Attempt to Map All the Genes in the Human Body*. New York: Simon and Schuster, 1990.

Bleifield, Maurice. *Experimenting with a Microscope*. New York: Watts, 1988.

Brown, Sheldon. *Opportunities in Biotechnology Careers*. Rev. ed. Lincolnwood, IL: VGM Career Horizons, 1994.

Gay, Kathlyn. *Cleaning Nature Naturally*. New York: Walker, 1991.

Glaser, Linda. *Compost: Growing Gardens from Your Garbage*. Brookfield, CT: Millbrook Press, 1996.

Hoffmann, M. P. and A. C. Frodsham. *Natural Enemies of Vegetable Insect Pests*. Ithaca, NY: Cooperative Extension, Cornell University, 1993.

Jackson, John. *Genetics and You*. Totowa, NJ: Humana Press, 1966.

Kornberg, Arthur. *The Golden Helix: Inside Biotechnology Ventures*. Sausalito, CA: University Science Books, 1995.

Lampton, Christopher. *DNA Fingerprinting*. New York: Watts, 1991.

Lavies, Bianca. *Compost Critters*. New York: Dutton, 1993.

Lee, Thomas. *Gene Future: The Promise and Perils of the New Biology.* New York: Plenum, 1993.

Mather, Robin. *Garden of Unearthly Delights: Bioengineering and the Future of Food.* New York: Penguin, 1995.

McGee, Harold. *On Food and Cooking. The Science and Lore of the Kitchen.* New York: Harper Collins, 1992.

Newton, David. *The Ozone Dilemma: A Reference Handbook.* Santa Barbara, CA: ABC-Clio, 1995.

Pringle, Lawrence. *Vanishing Zone: Protecting Earth from Ultraviolet Radiation.* New York: Morrow, 1995.

Rainis, Kenneth and Bruce Russell. *Guide to Microlife.* Danbury, CT: Watts, 1996.

Sherrow, Victoria. *James Watson and Francis Crick: Decoding the Secrets of DNA.* Woodbridge, CT: Blackbirch Press, 1995.

ORGANIZATIONS

Agricultural Biotechnology
National FFA Foundation
5632 Mt. Vernon Highway, P.O. Box 15160
Alexandria, VA 22309
Phone: (703) 360-3600

American Academy of Forensic Sciences (AAFS)
P.O. Box 669, 410 North 21st Street, Suite 203
Colorado Springs, CO 80901
Phone: (719) 636-1100
FAX: (719) 636-1993
e-mail: Internet membership@aafs.org

American Petroleum Institute
2101 L Street NW, Washington, DC 20037
Phone: (202) 820-8000

American Society of Human Genetics
P.O. Box 6015, Rockville, MD 20850
Phone: (301) 424-4120

Biotechnology Education Resources Guide
University of Wisconsin Biotechnology Center
Phone: (608) 265-2420
e-mail: zinnen@macc.wisc.edu.

Biotechnology Industry Organization (BIO)
1625 K Street NW, Washington, DC 20006
Phone: (202) 857-0244
FAX: (202) 857-0237

Genetics Society of America
9650 Rockville Pike, Bethesda, MD 20814-3998
Phone: (301) 571-1825

National Agricultural Library
10301 Baltimore Blvd, Beltsville, MD 20705-2351
Phone: (301) 504-5947
FAX: (301) 504-7098
e-mail: biotech@nalusda.gov

INTERNET SITES

Biotechnology Companies
To obtain a listing of many biotechnology companies and their Internet addresses, try
http://155.41.115.114/jumber/biocomm.html

Biotechnology Experiments
http://exploration.jsc.nasa.gov/explore/Data/Apollo/Part2?El
ecA14.htm

Biotechnology Information Center at the National Agricultural Library
http://www.inform.umd.edu/EdRes/Topic/AgrEnv/biotech

Biotechnology Public Education Program
http://biotech.zool.iastate.edu/Biotech_Public_Ed.html

Biotechnology Today
Biotech news and job information source
http://www.biotech.mond.org

Biotechnology Virtual Library
htmhttp://www.webpress.net/interweb/cato/biotech/

Center for Biotechnology Policy and Ethics
http://www.tamu.edu/cbpe

The Genome Database at Johns Hopkins University School of Medicine
http://gbwww.gdb.org/

Internet Directory of Biotechnology Resources
http://biotec.chem.indiana.edu

Karyotyping
Download prepared human karyotypes
http://www.selu.com/~bio/cyto.human/HSAhtml

Legal Issues and Biotechnology
Legal and scientific information on biotechnology patents
http://biotechlaw.ari.net
http://www.aphis.usda.gov

INDEX

extracted DNA, 45–46
extracting DNA,
  38–42, *41*
genetic engineering,
  51–53
paternity testing,
  133–134
Drain bacteria, 28
Drosophila. *See* Fruit fly

Electrophoresis apparatus,
  11–14, *12, 13*
building one, 14–18,
  15, 17
major components, 12
using it, 18–23, *20,*
  21, 22
Environment, 113
composting, 126–128
ice making, 116–118
oil spills, 122–126
radiation detector,
  113–116, *114*
staying ice-free,
  118–119
too much sun, 119–122
Enzymes, 110–112

Fermentation, 6, 7, 25.
  See also Bioreactors
fermentation media, 136
Flowers, 65
Fruit fly, 46–50, *47, 48*
working with, 137–138
Gel electrophoresis, 11–12.
  See also Electrophoresis
apparatus
Genetic disease, 8, 92–93.
  See also Genetic screening

DNA fingerprinting,
  131
Genetic engineering,
  51–53. *See also* Plants
Genetic screening,
  92–101, *94, 96, 100*
social impact of, 101
Growth phase, 28, *29*

Heat, and DNA, 42–45,
  43, 44
Herbicides, 66
Heredity, 48, 87–88. *See also*
Human genes
Human genes, 86–87,
  102–103, *103*
genetic screening,
  92–101, *94, 96, 100*
inherited characteristics,
  89
pedigree, 89–91, *90*
Punnett square, 91–92, *91*
Human Genome Project,
  102–103

Ice
making ice, 116–117
staying ice-free, 118–119
Industry, 104
cryopreservation,
  104–110, *109*
designer detergents,
  110–112
Insects, 67–68
staying ice-free,118–119
Insulin, 51
Internet sites, 157

Jeffreys, Alec, 130

40848

660   Rainis, Kenneth G.
RAI

Biotechnology
projects for young
scientists